A New Foundation for Chemistry

A New Foundation for Chemistry

Antoine Lavoisier's
Elementary Treatise on Chemistry
PRELIMINARY DISCOURSE AND PART ONE

Translated from the 1789 edition
by Chester Burke and Matthew Holtzman

With running commentary
by Howard J. Fisher

Green
Lion
Press

Manufactured in the United States of America.
Published by Green Lion Press
Santa Fe, New Mexico
www.greenlion.com

Green Lion books are printed on acid-free paper.

Set in 12-point Arno Pro
Printed and bound by McNaughton & Gunn, Saline, Michigan

Cataloguing-In-Publication data:
Lavoisier, Antoine
A New Foundation for Chemistry: Lavoisier's *Elementary Treatise*,
Preliminary Discourse and Part I.
Translated from the 1789 edition
by Chester Burke and Matthew Holtzman
with running commentary by Howard J. Fisher.
Includes index, glossary, and biographical note.

Cover: "Lavoisier in his Laboratory," hand-colored halftone of a drawing by Thule de Thulstrup.

Diagrams from the first 1789 edition, engraved by Marie-Anne Paulze Lavoisier. Courtesy of the History of Science Collections, University of Okalhoma, and the Science History Institute, Philadelphia.

Photo credits: p. 36—CNAM, Conservatoire National des Arts et Métiers, Paris/Bridgeman Images; p. 50—© Musée des arts et métiers–CNAM, Paris/photo J-C Wetzel.

ISBN-13: 978-1-888009-54-5 (sewn softcover binding)
1. Lavoisier, Elementary Treatise on Chemistry. 2. Chemistry.
3. History of Science.
I. Lavoisier, Antoine-Laurent (1743–1794). II. Title

Library of Congress Control Number: 2021950418

Table of Contents

Elementary Treatise on Chemistry

CONTENTS

The Green Lion's Preface

In preparing and refining the text and comments for this edition, the Green Lion editors and translators have been immeasurably aided by our long involvement, as students and teachers, in St. John's College's unique approach to the study of scientific texts. For over fifty years, on two campuses, every freshman has spent several weeks of the spring semester in a close study of Lavoisier's *Elementary Treatise*—a study that involves not only reading and discussion but also working through Lavoisier's chief experiments in the laboratory. The editors and translators have thus benefited from St. John's College's deep and rich body of experience that we can justly claim to be unequalled anywhere in the world.

Translators Chester Burke and Matthew Holtzman are faculty members at St. John's. In addition to an impressive command of the French of Lavoisier's time, they have many decades of experience reading Lavoisier's Treatise with nonspecialist undergraduate students. The Green Lion editors, Howard Fisher, Bill Donahue, and Dana Densmore, have all also had extensive experience reading and helping students read Lavoisier—all three of us starting in the 1960s. Fisher and Burke were important early and ongoing guides for the St. John's College laboratory program on the Annapolis campus, and Donahue was Director of Laboratories for many years on the Santa Fe campus.

Notwithstanding the admirable lucidity of Lavoisier's writing, his thought is premised upon materials and concepts that are likely to seem somewhat strange to nonspecialist readers—perhaps even stranger to readers boasting a modern chemical background. This Green Lion edition offers a running commentary that aims to respond to the text much as an experienced reading partner might do: explaining (but sometimes only speculating on) antiquated terminology, pointing out questions (but only sometimes answering them), and discussing the construction and use of Lavoisier's often original chemical apparatus. Howard Fisher has supplied these annotations.

The present volume is a partial realization of a long-standing goal of the Green Lion. We thought that Lavoisier's magnificent *Elementary Treatise* had too long been hobbled by an archaic English translation, and we hoped to set it free. Our aim always went far beyond modernizing the text. Because the present translation is different from other renderings of scientific works that readers may have encountered, we think it appropriate to say what we think a Green Lion translation should accomplish.

There is much variety among the Green Lion translations, but one feature they hold in common is a deep respect for the original text. We have all seen plenty of translations that are tinted, and sometimes tainted, by an evident feeling on the part of the translator that he or she knows the author's intent well enough to rephrase or even paraphrase passages that appear awkwardly or obscurely worded. These efforts certainly have their place, but that is not where we want to go. We are especially interested in the odd or clumsy expression, because those oddities often bespeak a remarkable and original idea for which the author does not yet have a clear expression. We want readers to be able to watch a profound thinker struggle to think through a new and difficult thought.

We are committed to a translation of this text that reveals and traces Lavoisier's thought and voice as much as possible. This means not eliding over nuances of diction, imagery, and metaphor that may bear precisely on the way Lavoisier was thinking about the phenomenon he describes. It means paying attention to maintaining Lavoisier's parlance, resisting paraphrase, and keeping a little closer to the structure of the French where there is not an overriding reason to depart from it. We value a translation that is scrupulous about bringing out the original author's written expressions, including images and subtle resonances, even at the expense of awkward and unfamiliar wording; this leaves room for students, readers, and scholars to dig out interesting angles. The translation should stand on the solid ground of the text, perhaps most of all when we are not sure why that ground has the shape that it displays. It should speak to those who study the poetry of Lavoisier, as well as those who study his science.

It has been our joint effort to bring through Lavoisier's thought in all its lucidity, preserving the excitement of this moment, a moment on the cusp of making new sense and order out of a jungle of chemical phenomena. The resulting translation is faithful to the text in the ways described above, while producing a version that reads transparently in English. We are honored that Burke and Holtzman have entrusted us with their work, and we are delighted to present it to a larger public, enhanced by Fisher's illuminating notes.

Dana Densmore
William H. Donahue
for the Green Lion

Acknowledgments

Illustrations in the text are the ones created for the first 1789 edition of the *Treatise* by Marie-Anne Pierrette Paulze—Mme. Lavoisier. These were generously provided by JoAnn Palmeri, Curator of the History of Science Collections at the University of Oklahoma, and carefully prepared for publication by William Donahue. In the few cases where the Oklahoma images cannot be clearly reproduced in print, we have used high-resolution photographs of the Paulze-Lavoisier engravings from a copy of the first edition in the collection of the Science History Institute in Philadelphia. These have been provided by James R. Voelkel, Curator of Rare Books at SHI, and are now available for download from the SHI web site. Illustrations in the comments are from various sources. Caroline Mason of the Los Alamos National Laboratory graciously shared a modern chemist's insight into the combustion of phosphorus, described in Chapter Five.

Annotator's Introduction

That knowledge becomes *science* only when it is organized in the form of laws—preferably, mathematical ones—has become almost a scientific credo; and when when writers laud Lavoisier as "the father of modern chemistry," it is invariably his introduction of quantitative thinking that they have in mind. But chemistry springs from a tradition that cultivated a very different conception of knowledge: knowledge as *direct apprehension* more than subordination to demonstrative logic. And Lavoisier's new science, revolutionary though it was, did not lose sight of that heritage.

Reading Lavoisier's *Elementary Treatise on Chemistry*, we feel his abiding awareness that *matter has powers*; and we can participate in his desire to know those powers firsthand. Laws—especially laws aiming for universality—are largely beside the point; for Lavoisier's awareness is far more attuned to the distinctiveness of different kinds and classes of substances.

For that reason it is rewarding to pay attention to Lavoisier's verbs. When does he use terms like "release" (*dégager*), as distinguished from "remove" (*enlevèr, retirer*) or "extract" (*extraire*)? Lavoisier seems constantly trying to infer exactly what each substance is doing in relation to the other substances in each chemical reaction. When he cannot deduce, he *conjectures*—because his interest is in figuring out how each substance acts, more than in demonstrating how it conforms to an overall pattern. To be sure, he loves those large patterns when he finds them; but they are always welcomed as gifts, as prized fruits of study and labor—never as the substance of knowledge itself.

Studying Lavoisier, like studying the early works of any science, will be most rewarding if we can temporarily set aside concepts that our schooling offered to us ready-made, and let them instead become live questions. By doing so we can appreciate, and even take part in, the working-out of those concepts by the great founding thinkers.

Consider, for example, Lavoisier's understanding of heat as being a *substance*, which he calls "caloric"—a view which modern textbooks too often treat with condescension or even ridicule. Heat, as we all now know, is no substance, but rather a form of energy. But readers who can step back from "what we all now know" are regularly astonished to discover just how strong was Lavoisier's evidence for the existence of caloric, and what a rich explanatory power that concept had in Lavoisier's hands. It does no honor to later investigators if we think the theories they superseded were merely silly. If we really want to appreciate the modern concept of heat, we will want to understand the potency and utility of the theory it supplanted.

Along with ready-made concepts, we will do well to forego the standardized *terminology* in which those concepts are embedded. This is particularly appropriate for Lavoisier, as *names* were of central importance to him. The "Preliminary Discourse" which begins this volume is his compelling call to name substances according to *what we actually know* about them, not just what convention dictates or what practice has ordained.

As knowledge becomes more and more widely accepted, technical terms tend to migrate into ordinary speech, even to the point where we may forget that they *are* technical terms—and that they therefore embody a host of assumptions and suppositions that were once open questions. We, to a far greater extent than Lavoisier's contemporary readers, need to be aware of this.

Consider the term "molecule," for example, which for us denotes a definite cluster of fundamental substantial units. But Lavoisier's *molécules* are, literally, no more than *small masses* of the substance in question; he has no reason to suppose that these bits of matter are equal to one another, even in the same substance. Such intimate details of the constitution of bodies appear to be out of reach of Lavoisier's methods; but readers are invariably surprised and delighted to discover how much one *can* know about materials, and how interesting they can be, even when their minute constitution remains a mystery.

Commensurate with Lavoisier's striking scrutiny of the vocabulary of chemistry was the scrupulous attention he devoted to its physical apparatus. Chemistry, like other sciences, cultivates a highly disciplined brand of experience that requires highly specialized equipment to produce. Lavoisier worked at a time when not only the basic concepts of chemistry were in flux, but even its instruments and procedures were not standardized. While certain articles of glassware—retorts, flasks, and the like—had long been associated with chemical and alchemical work, even these had to be made more or less on an as-needed basis rather than produced as stock items. The beam balance had long been used in trade and commerce, but Lavoisier made it an indispensable instrument for chemistry by demonstrating the need for greater precision than ordinary instruments could deliver. The exceptionally capable Fortin balance (page 50), which he commissioned, brought that familiar appliance to new standards. On the other hand, instruments like Lavoisier's gasometer (page 90) were wholly new.

To what degree of knowledge can chemistry realistically aspire? In contrast to the demonstrative paradigm, Lavoisier knows that conclusive answers are not the only valuable ones. His science is not a "method" that guarantees truth; instead, as we shall see, it generates *experience* that can either clarify or overthrow predominating conceptions.

This Green Lion edition makes use of a few formal devices that may need explanation. I have supplied paragraph numbering, in part to facilitate cross-references. Thus, for example, the number 3.10 in parentheses— (3.10)—directs the reader to consult paragraph 10 in Chapter Three for supporting remarks or other material related to the topic under immediate discussion. Pagination in the 1789 edition is indicated in the margins.

I have endeavored scrupulously to distinguish between Lavoisier's narrative and my commentary. Within each Chapter, his text always occupies the top of the page. My comments appear in smaller type at the bottom of the page, beneath a separator line, and keyed to each numbered paragraph— in most cases also to a particular phrase. Lavoisier's own footnotes appear beneath his text, but above the separator line. A few translators' notes are also placed above the separator line.

Howard J. Fisher
December 2021

About Antoine Lavoisier

Antoine-Laurent Lavoisier was born in 1743 and began his formal education at age 11 at the Collège des Quatre Nations in Paris, where he received a sound classical and literary training as well as, perhaps, the best scientific education then available.

First active in geology and mineralogy, his endeavors led Lavoisier to attend the chemistry lectures of Guillame-François Rouelle in 1762 and 1763. His earliest chemical investigation—a study of gypsum in 1764—was indeed mineralogical in character. Following his presentation of a major paper on hydrometry in 1768, he was elected to the Academy of Sciences despite his relative youth.

In the 1770s and 80s, when Lavoisier's scientific activity was at its peak, his research included the study of gases, vaporization, heat, respiration, and the formation of acids. His most dramatic triumphs were a series of experiments in 1784–5 establishing the constitution of water (thereby proving that it was not an element) and his 1786 refutation of the phlogiston theory of combustion. A new chemistry was rapidly emerging.

Lavoisier himself had envisioned nothing less than "a revolution in physics and chemistry" in 1773, the founding document of which might be said to be the *Méthode de nomenclature chimique* of 1787, a joint work with Morveau, Berthollet and Fourcroy which undertook to make the language of chemistry reflect the nature of the substances it named. But the most influential publication promoting the new thinking was undoubtably Lavoisier's *Elementary Treatise on Chemistry,* which appeared in 1789. More than a systemization of the science—which would itself have counted as a major achievement—this work also contained new, previously unpublished research on such topics as the constitution of organic substances.

In 1771 Lavoisier had married the young but very gifted Marie-Anne-Pierette Paulze. The marriage became a fruitful scientific collaboration. She was to become deeply involved in Lavoisier's scientific activities, assisting in the laboratory, documenting experiments and apparatus in sketches and notes, and translating the scientific writings of English researchers. She recorded the progress of numerous experiments, especially those on respiration. She was trained as an artist and engraver (she studied with the painter

Jacques-Louis David) and it was she who prepared the fine engravings for the 1789 publication of the *Elementary Treatise on Chemistry*—engravings that are reproduced in this Green Lion edition. Both during his life and after his death, she acted as an energetic promoter of her husband's thinking and reputation.

Lavoisier's scientific interests grew largely from private curiosity, but his concerns embraced the public welfare to at least an equal extent. He addressed such problems as securing an adequate water supply for Paris, inadequate street lighting, the deplorable conditions in prisons and hospitals, and advancing improvements in agriculture and livestock.

Politically liberal, though holding moderate views, Lavoisier was active in several of the efforts that led to the Revolution. He served on a standing committee concerned with conditions in agriculture and trade, where he reported on charitable foundations and proposed systems of old-age insurance and tax reform. But, notwithstanding his many services to his country and to science, Lavoisier increasingly became the target of vicious attacks in the growing extremist atmosphere, in whose spirit to be moderate was to be traitorous. In 1768 Lavoisier had purchased a share in a company that collected customs, excise, and other taxes under royal charter. Although the income from this source helped finance his scientific work, Lavoisier's association with tax collection later made him a target of the revolutionary power. On 8 May 1794 Lavoisier was executed by the Revolutionary Tribunal at age 50.

The mathematician Lagrange memorialized Lavoisier's execution with these words: "It took them but an instant to cut off this head, and one hundred years might not suffice to produce another like it." A year and a half after his death, Lavoisier was exonerated by the French government.

ELEMENTARY TREATISE ON CHEMISTRY

Preliminary Discourse

Preliminary Discourse

0.1 When I first undertook this volume, my only aim was to develop further the paper I read at the public meeting of the Academy of Sciences in April, 1787 on the necessity of reforming and perfecting the nomenclature for chemistry.* [*v*]

0.2 But while I was engaged in this work, I came to appreciate, better than I ever had before, evidence for the principles that the Abbé de Condillac set out in his *Logic* and in other works. Condillac has established that "we only think with the assistance of words," that "languages are true analytical methods," that "algebra, being the simplest, the most precise, and the most suitable to its object among all modes of expression, is both a language and an analytical method"; and, finally, that "the art of reasoning is, at bottom, [*vi*] nothing else but a well-constructed language." And so it turned out that while I thought I was only focused on nomenclature, and while my sole intention was to perfect the language of chemistry, my work, in a manner I could not resist, imperceptibly transformed in my hands into an elementary treatise on chemistry.

0.3 The impossibility of isolating nomenclature from science and science from nomenclature arises because every physical science is necessarily composed of three things: the series of facts which constitutes the science; the ideas that recall those facts; and the words that express those ideas. Now the word should give rise to the idea; the idea should depict the fact. Word, idea and fact are three imprints of the same seal; and because words preserve

---

* [Lavoisier refers to the first section (pp. 1–25) of *Méthode de nomenclature chimique* (Paris, 1787), which he authored jointly with de Morveau, Berthollet, and de Fourcroy. —Tr.]

0.1 *the paper I read ... in April, 1787*: Lavoisier refers to the first section (pp. 1–25) of *Méthode de nomenclature chimique* (Paris, 1787), which he authored jointly with de Morveau, Berthollet, and de Fourcroy.

0.2 *the Abbé de Condillac set out in his Logic*: Condillac, *La logique* (Paris, 1780).

and transmit ideas, one cannot perfect the language without perfecting the science, nor the science without the language. And no matter how certain the facts, no matter how sound the ideas that the facts produce, they would convey nothing but false impressions were we not in possession of exact expressions to render them.

[vii]

0.4 The first part of this treatise will provide those who carefully consider it with many demonstrations of these truths, but since I have been obliged to follow a plan for the treatise which differs essentially from the one currently adopted in all works of chemistry, I should explain the reasons that led me to do so.

0.5 It is a well-established principle, a principle recognized as quite generally applicable in mathematics and, indeed, in all branches of knowledge, that instruction can only proceed from the known to the unknown. In early childhood, ideas arise from needs. The perception of our needs produces an idea of objects that could satisfy them, and imperceptibly one idea after the next arises from our sensations, observations, and analyses, and all of these ideas are tied to one other in such a way that an attentive observer could, up to a certain point, even recover the thread that connects them. This chain of thoughts constitutes the sum of what we know.

[viii]

0.6 We are much like children with respect to any science we have taken up for the first time, and we must follow the very path that nature follows in forming a child's ideas. For a child, ideas are the effects of sensations, sensations produce ideas, and the same should be true for all those who devote themselves to the study of the physical sciences: their ideas ought to be the consequence, the immediate result, of an experiment or observation.

0.7 I would like to add that whoever embarks on a career in the sciences is in an even less advantageous situation than the child acquiring its first ideas. If the child is mistaken about the helpful or harmful effects of objects in its environment, nature provides it with ample opportunity for correction. At each moment, experience amends a child's judgment. Privation and pain follow false judgments; joy and pleasure follow sound ones. It does not take long to become a very respectable thinker with teachers like these, for one quickly learns to reason well when the penalty for reasoning badly is pain or hardship.

[ix]

0.6 *ideas ought to be the consequence ... of an experiment or observation*: Knowledge in the sciences should begin with *actual experience*—not with, say, doctrines inherited from tradition, or (as he will discuss in the next paragraph) with ideas we just come up with on our own.

0.8 But this is not at all true in the case of the study or the practice of science. The false judgments we make have no bearing on our existence or well-being, and no physical concerns oblige us to revise them. On the contrary, imagination (which constantly tends to carry us beyond the truth), as well as self-love and self-confidence (which all too often motivate us), tempt us to draw conclusions which by no means follow immediately from the facts, so we are in a sense involved in deceiving ourselves. [x]

0.9 It is hardly astonishing, then, that throughout the physical sciences suppositions so often take the place of conclusions, that these suppositions, handed down from one age to the next, should grow increasingly imposing as they acquire the weight of authority, or that they should, in the end, be adopted and regarded as fundamental truths by even the greatest minds.

0.10 There is only one method for preventing errors like these. We must eliminate, or at least simplify, our inferences whenever possible, since we are alone responsible for the inferences we make, and so they are the only things which can mislead us. We must constantly refer our inferences back to the touchstone of experience. We must only retain a record of the facts, since facts are nothing but the things nature gives us and so they cannot deceive us. And we must search for truth only among the natural connections between experiments and observations, like mathematicians when they are working towards the solution to a problem. For mathematicians merely arrange the givens and reduce all lines of thought to such simple steps and such straight- [xi] forward judgments, that they never lose sight of the evidence that guides them.

0.11 Convinced by these truths, I adopted the following rules: to proceed only from what is known to what is unknown, to draw only such consequences as fall out immediately from experiments and observation, and to arrange the truths of chemistry in the order most appropriate for facilitating the understanding of beginners.

0.12 It was impossible for me to pursue this plan without departing from standard approaches. For almost every chemistry course and treatise makes

---

0.8 *false judgments [in science] have no bearing on our existence or well-being*: False ideas about everyday matters are swiftly corrected by casual experience. But the experience that science cultivates is decidedly restricted and often highly contrived. An erroneous opinion concerning the result of touching a flame will be rectified at once; but a false opinion about whether water is a simple or a compound substance does not affect everyday life and is unlikely to be corrected without the construction and use of elaborate specialized apparatus.

the mistake of presupposing at the outset knowledge that students or readers should only acquire much later in their studies. They almost all begin by treating the elements of bodies and explaining the table of affinities without realizing that this requires them, from the very first day, to present the fundamental phenomena of chemistry, to use expressions which have not yet been defined, and to assume that the student or reader has already acquired the very knowledge that the treatise or course is meant to impart. And so it is acknowledged that little can be learned in a first course in chemistry; that a year of study is hardly enough to develop an ear for the language of chemistry or an eye for its instruments; and that it is almost impossible to train a chemist in fewer than three or four years.

0.13 But these difficulties pertain not so much to the nature of the subject as to the manner in which it is taught; and this is what determined me to set chemistry on a course that seemed to me more in accordance with nature. I was aware that in order to avoid one kind of difficulty I have fallen into others,

---

0.12 *the mistake of presupposing at the outset knowledge that students or readers should only acquire much later*: Aristotle distinguished between *the order of being*, which articulates what matters are primary or secondary in themselves, and *the order of knowing*, which identifies what matters are more and less immediately accessible to us. Lavoisier has been arguing, in effect, that learning must proceed according to the order of knowing; but traditional chemistry treatises have endeavored to present their material in a sequence that they consider to reflect the order of being. Not only does this constitute poor educational practice in Lavoisier's view, but it distorts the character of science, which—since it proceeds from the known to the unknown (0.5)—must necessarily reflect the order of knowing.

the elements of bodies: For example, the traditionally-accepted elements earth, air, water, and fire. To begin by discussing these is indeed an example of presupposing at the outset knowledge that can only be acquired later. For as he will note in (0.21), the determination that a particular substance is "simple" requires continual experimentation and, in a way, can never be regarded as altogether complete.

the table of affinities: The tendency of one substance to combine with another is called its *affinity* for that substance. It was customary for treatises on chemistry to present this information in the form of tables such as Geoffroy's of 1718, a portion of which is shown here.

TABLE DE Mᴬ GEOFFROY en 1718.

an ear for the language ... an eye for its instruments: Here Lavoisier seems to express the essential skills of the chemist; for the *instruments* are the means of gaining the experience on which chemical knowledge is based (0.6), while the *language* is practically synonymous with the organized body of knowledge itself (0.3).

and that it would be impossible to avoid them all. But I believe that those which remain may be ascribed to imperfections in the science of chemistry rather than to the direction I have prescribed for myself. It is difficult and confusing to find the connections between the facts of chemistry, because the series of chemical facts is riddled with gaps. Unlike elementary geometry, [*xiii*] chemistry lacks the advantage of being a complete science in which all parts are closely bound to one another. But at the same time, it is currently progressing so quickly, and facts are organized so felicitously under the modern doctrine, that we can hope, even in our own time, to see chemistry approach the highest degree of perfection it can reach.

0.14 These strict rules, from which I have never deviated, and which keep me from drawing any conclusions aside from those experiments show me, or from speaking when the facts are silent, did not permit me to consider in this work the branch of chemistry most likely, perhaps, to become an exact science someday: the branch devoted to chemical affinities or elective attractions. Geoffroy, Gellert, Bergman, Scheele, de Morveau, Kirwan and many others have already assembled a vast array of particular facts and it only [*xiv*] remains for someone to assign these facts their proper places. But we lack the principal data, or at least the data we have are still not accurate or certain enough to serve as a foundation for such an important branch of chemistry. Besides, the science of affinities is to ordinary chemistry what transcendental geometry is to elementary geometry, and I do not think that we ought to complicate the elements of chemistry, which are simple, easy and, I hope, accessible to a great many readers, with grave difficulties.

0.15 Perhaps a sense of pride has, without my realizing it, motivated these reflections. De Morveau is just about to publish an article entitled "Affinity" in *l'Encyclopédie méthodique*, and I have many good reasons to dread competition with him.

0.16 It will be surprising not to find a chapter on the elementary and consti- [*xv*] tutive parts of bodies anywhere in this elementary treatise on chemistry, but I

0.14 *elective attractions*: an alternative name for *affinities*, expressing a substance's tendency to combine with certain substances in preference to others; from "election" as *exercise of choice according to preference*.

transcendental geometry: The geometry of transcendental curves.

0.16 *the elementary and constitutive parts of bodies*: A strong tradition, dating back to Aristotle, dictated that treatises on chemistry should identify a number of "elementary" substances, from which all other substances were supposed to be formed. Aristotle had singled out *earth*, *air*, *water*, and *fire* (representing, in turn, the four possible pairs of *moist* or *dry* and *warm* or *cold*). The 16th-century alchemist Paracelsus identified *sulphur*, *mercury*, and *salt* as elements.

will say here that we are attracted to the view that all bodies are composed of three or four elements because of a prejudice inherited from Greek philosophers. The supposition that four elements, in various proportions, compose all known bodies is a pure hypothesis conceived long before anyone had the first notion of experimental natural philosophy and chemistry.

0.17 The Greek philosophers had no facts and devised systems, and it is a testament to the weight of their authority that today we do our best to reject the facts we have gathered since their time whenever these facts cannot be squared with our prejudices, and there can be no doubt that their authority [*xvi*] will continue to weigh on generations to come.

0.18 But it is well worth noting that every chemist who has taught the doctrine of the four elements has been forced by the facts to admit more than four. Chemists writing just after the Renaissance regarded sulphur and salt as elementary substances involved in the composition of many bodies, so they recognized the existence of six elements, not four.

0.19 Becher admitted three kinds of earth, and he believed that differences between metallic substances depended on the proportions of these earths when they combined. Stahl modified this system, and all the chemists who succeeded Stahl took the liberty of making changes and imagining further changes, but all of them allowed themselves to be swept up in the spirit of an age content with unproven assertions or, at least, content to take the slightest [*xvii*] probabilities for proofs.

0.20 As I see it, all discussions of the number and the nature of the elements must be considered purely metaphysical because they involve the resolution of underdetermined problems, problems admitting of infinite solutions, none of which, in all likelihood, are in accordance with nature. So I will only say that if we mean, by "elements," the simple and indivisible particles that compose bodies, then it is likely that we are unacquainted with any elements. But if, on the other hand, we attach the terms "elements" or "principles" to the idea of the last result our analyses of bodies have obtained, all substances

---

0.20 *the simple and indivisible particles that compose bodies*: the actual constituents from which bodies are built up, and which are not, in turn, composed of other, simpler, particles.

it is likely that we are unacquainted with any elements: Chemistry has no means with which to bring forth the smallest possible particles of any substance; so if that is what "element" means, we have probably never experienced any elements— certainly not knowingly.

which we have not by any means been able to decompose will be elements for us.

0.21 We cannot guarantee that the bodies we regard as simple are not themselves composed of two or even a great number of principles. However, if the parts of a body never separate, or, rather, if we can find no way to sep- [*xvii*] arate them, we should regard that body as simple and should not consider it composite until experiment and observation supply proof.

0.22 These reflections on the way ideas should develop apply naturally to the choice of words for expressing those ideas as well. Guided by the work that de Morveau, Berthollet, de Fourcroy and I produced together in 1787 on chemical nomenclature, I have designated simple substances with simple words whenever possible, and I was obliged to name them first. As you may recall, we strove to retain all common or customary names for substances. We only permitted ourselves to change a name in two cases: first, in the case of newly discovered substances which had either not yet been named, [*xix*] or had been named so recently that the names had not yet gained common currency; and, second, in the case of terms, whether ancient or modern, which seemed to us to suggest obviously false ideas, as when, for instance, a term could cause confusion between its referent and another substance endowed with different or even contrary properties. In cases like these we felt no qualms about substituting other terms, usually derived from Greek, which express the most common and characteristic property of the substances. For beginners, this approach to naming has two advantages. It eases the burden of memorizing names, since it can be difficult for beginners to retain new words that are completely devoid of sense. It also accustoms them from the outset to admit into their vocabularies only those words that are attached to ideas.

0.23 We have designated bodies formed from the combination of many simple substances by names compounded in the very same way. But since [*xx*]

0.21 *We cannot be sure that the bodies we regard as simple are not themselves composed...*: Because a substance that has withstood decomposition up to now may someday yield to a hitherto undiscovered technique.

0.22 *I have designated simple substances with simple words*: This principle constitutes the beginning of Lavoisier's effort to *perfect the science* by *perfecting its language* (0.3).

terms ... which ... suggest obviously false ideas: In paragraph (0.33) below, Lavoisier will single out for scorn names like "flowers of zinc" and "butter of arsenic" on this account.

even the number of binary compounds is already quite considerable, we would have plunged into disorder and confusion if we had not committed ourselves to construct names in such a way as to form classes.

0.24 In the natural order of ideas, the word for a class or genus is the word that calls to mind some property common to a great many individuals: the word for a species, on the other hand, is that which brings to mind the idea of the properties of a few.

0.25 This is not a mere metaphysical distinction as one might suppose: it is, rather, natural. A child, Condillac tells us, will call the first tree we show him "tree." When the child sees a second tree or a third or fourth, he recalls the same idea and gives each of them the same name. In this way, the word "tree," which he first used to refer to an individual, becomes for him a name [xxi] for a class or a genus, for an abstract idea that comprehends all trees in general. But once we have pointed out to him that all trees do not serve the same purposes, or bear the same fruits, he will soon learn to distinguish them from one another by separate, specific names. This is the logic of every science, and it naturally applies to chemistry.

0.26 Acids, for example, are composed of two substances that we classify as simple. The first constitutes an acid's acidity and is common to all acids, and

---

0.23 *binary compounds*: Substances that are composed of two constituents.

construct names in such a way as to form classes: When one substance is a constituent of several compounds, those compounds form a class. Lavoisier will name the substances of each class in such a way that their names will indicate the constituent they have in common, and thus signal the class to which they belong.

0.24 *genus ... species*: The plural of *genus* is *genera*. Today these terms are most commonly encountered in the system of classification set up by Carl Linnaeus in 1735, in which natural beings were grouped into three kingdoms (animal, vegetable, mineral), each divided into classes, and these in turn into orders, genera, and species, successively. (In some cases, Linnaeus recognized a rank even lower than the species.)

0.25 *This is the logic of every science, and it naturally applies to chemistry*: As Linnaeus named each natural organism according to its genus and species, Lavoisier will form his chemical names—at least for binary compounds—in the same fashion; the following paragraphs offer some fascinating examples.

0.26 *Acids*: The French adjective *acide*, which means *tasting sour or bitter*, goes back to 1545. One of the properties, then, which acids have in common is a sour taste. *Vinegar* seems to have been the paradigm of acidic taste; but the herb *sorrel* was another commonly-cited example. Vitriolic acid, associated with sulphur, exhibited vigorous destructive properties and so pointed to other "acidic" properties going beyond taste.

the name for the class or genus should be derived from this substance. The second is the substance peculiar to each acid, that which differentiates acids from one another, and it is from this second substance that the name of the species should be derived.

0.27 But in most acids, the two constitutive principles, the acidifying principle and the principle acidified, can exist in all the proportions that determine their points of equilibrium or saturation. This is what we observe in sulphuric acid and sulphurous acid, and we have expressed these two states of the same acid by varying the ending of the name for the species.

[*xxii*]

0.28 Metallic substances which have been exposed to the joint powers of air and fire lose their metallic sheen, gain weight, and take on an earthy appearance. In this state they are composed, like acids, of a single principle common to all of them and a particular principle proper to each. So it was necessary to classify these metallic substances in the same way as acids, by first grouping them under a name derived from the principle common to all of them. We adopted the word *oxide* for this purpose. We then differentiated them from one another by names denoting particular kinds of metals.

---

the first constitutes the acid's acidity: Lavoisier will describe the discovery and naming of this "acid-making" substance in Chapter 4 and its role in the formation of acids in Chapter 5.

the name for the class or genus should be derived from this substance: Lavoisier's chemical names have a significance that Linnaeus' nomenclature never even aspired to. The common daisy, *bellis perennis*, belongs to the genus *bellis* (from Latin *bellus*, pretty). But no one thinks that every member of that genus contains a unique substance that constitutes its beauty. For acids, however, this *is* the case; moreover, the second of its two constituents is the basis of those properties that distinguish any particular acid from other acids.

0.27 *the two constitutive principles can exist in the proportions that determine their points of equilibrium*: For example, sulphuric acid and sulphurous acid are composed of the same two constituents, but in different proportions. Lavoisier will discuss this variability, and the naming practices that express it, in Chapter 6.

0.28 *Metallic substances ... lose their metallic sheen, gain weight, and take on an earthy appearance*: Here is another class of substances which display similar characteristics. When metals are roasted in air they acquire the properties Lavoisier has listed and become what early chemists called *calces* (singular *calx*). In Chapter 7 Lavoisier will argue that they all share a common constituent—the same one, he claims, that acids do.

we adopted the word oxide: He will replace the old name "calx" with "oxide," a term that derives from the name of the shared constituent.

0.29 The combustible substances that are the specific and particular principles of acids and metallic oxides can be the common principles of a great many other substances as well. For a long time, the only known compounds of this class were compounds of sulphur, but we know today, after the experiments of Vandermonde, Monge and Berthollet, that charcoal combines with iron and, perhaps, with many other metals, forming steel, plumbago, and so on, depending on the proportion of charcoal. We also know, after the experiments of Pelletier, that phosphorus combines with a great many metallic substances. So we have also classified these different compounds using general names derived from a common substance, with endings that will help remind us of this analogy. We then differentiated them from one another with names derived from the substances proper to each.

[xxiii]

0.30 The nomenclature for things composed of three simple substances presented a few more difficulties because there are so many of them, and, above all, because it is impossible to express the nature of their constitutive principles without using highly compounded names. In the case of the bodies that are members of this class, such as the neutral salts, it was necessary to consider, first of all, the acidifying principle common to them all, second, the acidifiable principle that makes them acids, and third, the saline, earthy or metallic base which determines the particular species of salt. We have borrowed the name of each class of salts from the acidifiable principle common to all members of the class. We have then distinguished each species by the name of the saline, earthy, or metallic base peculiar to each.

[xxiv]

0.31 However, a salt composed of the same three principles can be in very different states, simply due to the proportions between those three principles. The nomenclature that we have adopted would be defective if it had been unable to express these different states, and we have achieved this by systematically changing the endings of the names of different salts in the same state.

0.29 *combustible substances*: In common usage, *combustible* materials are those capable of being burnt or consumed by fire. Lavoisier will show that *burning* is a process in which a substance, or one of its constituents, combines with the common principle of acids and oxides.

plumbago: Here, Lavoisier lists it as one of several combinations of charcoal and iron, in keeping with the name *carburet of iron*, which he and his co-authors had introduced earlier in the *Méthode de nomenclature chimique* (Paris, 1787). But "plumbago" was a name far more commonly applied to any of several ores of lead (Latin *plumbum*), or to "black lead," which is not lead at all; see the glossary entries for *plumbago* and *graphite*.

0.32 So at last we have achieved a perspective from which it is possible to   [*xxv*]
recognize a number of different properties of a compound just by looking
at its name. We can see which substance in the compound is combustible,
whether that combustible substance is combined with an acidifying principle, and, if so, in what proportion. We can also tell at a glance what state an
acid is in, with which base it is united, whether the saturation is complete,
and, if not, whether the acid or base is in excess.

0.33 Clearly, it was not possible to take such different approaches to naming
without doing violence to customary usage from time to time, and without
adopting terminology that seemed harsh and barbarous at first. But in our
experience one quickly develops an ear for new words, especially when they
are part of a universal, rational system. Furthermore, the names used up until
now, such as "powder of algaroth," "salt of alembroth," "pompholix," "phagedenic water," "turpeth mineral," "colcothar," and many others are no less
harsh, no less strange. It requires long practice and a prodigious memory to   [*xxvi*]
call to mind the substances these words express, let alone to remember the
class of compound to which they belong. The names "oil of tartar per deli-
quium," "oil of vitriol," "butter of arsenic," "butter of antimony," "flowers of
zinc," and so on are still more improper insofar as they give rise to false ideas.

0.32 *to recognize ... properties of a compound just by looking at its name*: The advantages
of such names go beyond mere convenience, though the convenience is considerable.
In paragraph (0.22) above, Lavoisier recommended that a scientific vocabulary
should include "only those words that are attached to ideas." Names like "butter of
antimony" are attached to superficial ideas, based on a substance's appearance; but
Lavoisier's names are, ideally, based on the actual constitution of each substance and
so express knowledge of a type both more profound and more useful.

with which base it is united: Present-day readers may have been taught to think
of *bases* as specifically opposed to *acids*. But Lavoisier uses the term "base" more
generically: as the *foundation*, so to speak, of a more complex substance. For example,
he regularly speaks of the "base" of a gas—by which he means that substance which,
when combined with a sufficient quantity of heat ("caloric"), becomes the gas in
question. Elsewhere he writes of "an acidifiable base," meaning a combustible
substance which, combining with the universal acid-making substance, forms a
specific acid.

In Chapter 5 he will write: "bases capable of forming neutral salts." While it is
indeed *acids* with which such bases must combine in order to form salts, Lavoisier's
phraseology presents these bases as the *foundations of salts*, rather than the *antitheses
of acids*. For Lavoisier, a "base" is always the base *of* something, not a definite nature
in itself.

For, properly speaking, there are no such things as butters, oils or flowers in the realm of minerals, much less in the realm of metals, and the substances to which these misleading names refer are, after all, powerful poisons.

0.34 We were criticized when we published our *Essay on Chemical Nomenclature* for having changed the language our teachers spoke, the language that they had explained and passed on to us. But our critics forgot that Bergman and Macquer themselves urged this reform. The learned professor [*xxvii*] of Uppsala, Bergman, wrote to de Morveau towards the end of his life:

> You must never condone unsuitable terminology: those who are already well-versed in the subject will always understand you, those who are not will learn more quickly.

0.35 It would perhaps be more appropriate to criticize me for failing to give an historical account of my predecessors' views in a work that I am presenting to the public, that is, for only presenting my own opinions without discussing the views of others. Because of this I have not treated my colleagues, let alone foreign chemists, as justly as I had intended. But I beg the reader to consider that if one were to fill an elementary treatise with quotes, or deliver disquisitions on the history of science and the work of historically significant figures, one would lose sight of the true aim of such a treatise and would turn the work into a lecture that beginners would find thoroughly tedious. An ele- mentary treatise is neither a history of science nor a history of the human [*xxviii*] spirit. One should only aim in such a work for ease and clarity and should scrupulously avoid distractions. The path should be continuously smoothed over until no obstacle that could possibly impede someone remains. The sciences in themselves already present enough difficulties without adding more that are extraneous to them.

0.36 As for attributions, chemists, at least, will easily see that in the first part I only appeal to experiments I have conducted myself. If I sometimes adopt without proper citation the experiments or the opinions of Berthollet, Fourcroy, Laplace, Monge, or others who share my principles, it is because these chemists and I, by living together, and sharing our ideas, observations and ways of seeing things with one another, have established among our- [*xxix*] selves a community of thought in which it is often difficult, even for us, to distinguish precisely whose ideas are whose.

0.37 All that I have explained above concerning the order I have tried to follow in setting out proofs and ideas only applies to the first part of this work. This is the part that contains all of the doctrines that I have adopted, and this is the only part that I have sought to give a truly elementary form.

0.38 The second part is principally composed of tables of nomenclature for neutral salts. I have accompanied this only with very summary explanations of the simplest procedures for obtaining the different kinds of known acids.

0.39 This second part contains nothing that is properly my own. It merely presents a highly compressed abstract of results derived from the works of others.

0.40 Finally, in the third part, I have provided a detailed description of all modern chemical procedures. It seems that a work like this has been wanted [*xxx*] for a long time, and I believe it will be of some value. Generally, the use of experiments, especially modern experiments, is not sufficiently widespread, and if I had given more detailed instructions in the different memoirs I read to the Academy, perhaps I would have made myself more easily understood and science would have progressed more rapidly. The structure of the content of this third part seems to me almost arbitrary, and in each of its eight chapters my only goal was to group together the procedures which seemed most analogous to one another. It will be easy to see that this third part could not have been derived from any other work and that in the main sections I have been aided only by my own experience.

0.41 I will conclude this preliminary discourse by transcribing word for word some passages from Condillac which seem to me to represent faithfully the state of chemistry up until very recently.* These passages, which were [*xxxi*] written for another purpose, do but gain force when applied to the subject at hand.

> Rather than observing the things that we wished to understand, we preferred to imagine them. We became lost in a crowd of errors as we wandered from one false supposition to the next. These errors became prejudices, and so we took them for principles, leading ourselves ever further astray. Because of this, we acquired bad habits of reasoning, and eventually we could only reason in accordance with these habits. For us the art of reasoning became the art of abusing words without properly understanding them.... When we have reached such a pass, when errors have piled up to such an extent as this, there remains no other means of restoring order to the faculty of thought than to forget all that we have known, to take up our ideas again from their origins and follow out their [*xxxii*] development: to recast the human understanding, as Bacon puts it.

* Condillac, La logique, Part II, Chapter 1.

0.41 *forget all that we have known*: Our institutionalized "knowledge" has become so corrupted by unexamined presuppositions and hasty inferences that the only remedy is a radical reconstruction, to "take up our ideas again from their origins." These words, written in 1780, can hardly fail to suggest to a modern reader the revolutionary backlash against a corrupt and incorrigible monarchy—a sweeping away that would inflame all France within the decade and take Lavoisier's life in 1794.

The better educated one fashions oneself, the more difficult it will be to begin again. And so scientific works, which are very precise, clear, and well-ordered, will not be accessible to everyone. Those who have never studied will understand them more easily than those who have studied the sciences for a long time, and much more easily than those who have written a lot about them.

0.42 Condillac adds at the end of Chapter 5,

But the sciences have in fact progressed because natural philosophers have made better observations, and because they articulate them in a language as precise and exact as the experiments themselves. In doing so, they have corrected the language and have reasoned better.

0.41, cont. *Those who have never studied will understand them more easily than those who have studied the sciences for a long time*: Readers new to the study of science are urged to take heart from Condillac's words. Bear in mind that Lavoisier is undertaking to free chemical knowledge from the weight of long tradition, as well as from the inherent human tendencies towards facile reasoning. We in the present day live amidst an even greater store of brilliant scientific achievement—which must nevertheless be put aside, at least provisionally, if we wish to experience the foundations for ourselves. If you are a relative novice in the sciences, you will find this task much less onerous than if you had been thoroughly schooled in scientific theory.

ELEMENTARY TREATISE ON CHEMISTRY

Part I

On the Formation and Decomposition of Aeriform Fluids,
and on the Combustion of Simple Bodies
and the Formation of Acids

Chapter One
On Caloric Compounds and the Formation of Elastic, Aeriform Fluids

1.1 It is an invariable natural phenomenon, and, as Boerhaave has firmly established, a quite general one, that when any body is heated, whether it [2] is a solid or a fluid, it expands in every direction. Facts on which one has relied in order to limit the scope of this principle are misleading, or at least are complicated by unusual circumstances that interfere with them. But once we have isolated all effects and have assigned to them their proper causes, we realize that the separation of particles by heat is a constant and universal law of nature.

1.2 If we first heat a solid body and, by heating it, increasingly separate its particles from one another, and then we later allow it to cool, these same particles approach one another again, observing the same proportion according to which they separated. The body, at each degree of temperature as it contracts, is restored to the very same dimensions it possessed while it was expanding, and if it is restored to its original temperature, it resumes the same volume it had initially, as far as can be perceived. But since we are still quite far from possessing the means to obtain a degree of absolute cold, and since, so far as we know, there is no degree of coldness that we can assume cannot be increased, we have never yet brought a body's particles as close to one another as pos- [3] sible, which means that the particles of every body in nature do not touch each other. This is a very strange conclusion, but it cannot be denied.

1.3 It is clear that since heat continuously induces the particles of bodies to separate from one another, there are no connections between particles, and there would be no solid bodies at all if the particles were not restrained

---

1.1 *particles*: French *molécules*; but in contrast to the modern technical term "molecules," Lavoisier's "particles" are simply *small portions* of a substance, not necessarily either the smallest possible portions or even of uniform size.

1.2 *the body, at each degree of temperature ... will be restored to the very same dimensions*: A mercury thermometer assumes this principle by treating the volume of the confined fluid as the indicator of temperature.

by another force that tends to draw them together and, as it were, link them to one another. This force, whatever its cause may be, has been called "attraction."

1.4 And so the particles of bodies can be regarded as subject to two forces, one repulsive and one attractive, between which they are in equilibrium. Whenever the attractive force prevails, bodies remain in a solid state, but if, on the contrary, attraction is the weaker force, if heat separates the particles of the body from one another to such an extent that they no longer remain within the spheres of each other's attractive influence, they no longer adhere to one another and the body ceases to be a solid.

[4]   1.5 Water constantly presents us with examples of these phenomena. Below zero degrees of the French thermometer, water is in a solid state, and is called "ice." Above this temperature, particles of water are no longer restrained by reciprocal attraction for one another, and they become what is called a "liquid." Finally, above 80 degrees, the particles yield to the repulsive force of the heat. The water takes on the state of a vapor or gas, and converts into an aeriform fluid.

1.6 The same could be said of all bodies in nature. They are either solids, liquids, or in an elastic and aeriform state, according to the ratio that exists between the attractive force of their particles and the repulsive force of heat, or, what comes to the same thing, according to the degree of heat to which they have been exposed.

1.7 It is difficult to conceive of these phenomena without admitting that they are the effects of a real, material substance, an exceedingly subtle fluid that insinuates itself between the particles of bodies and separates them. And even supposing that the existence of this fluid were hypothetical, it explains natural phenomena quite felicitously as will be shown later.

1.8 Since this substance, whatever it might be, is the cause of heat, or, in
[5]   other words, since the sensation that we call heat is the effect of the accumulation of this substance, we cannot in precise language call the substance "heat." For the same term cannot express both cause and effect. It is this that

---

1.5 *the French thermometer*: A scale of temperature according to which water freezes at 0° and boils at 80°, as Lavoisier will state explicitly in paragraph 1.13. Today it is commonly called the Réaumur scale.

aeriform fluid: A substance must be either solid, liquid, or aeriform; it is aeriform when it has the air-like state exhibited by gases and vapors. A fluid is anything that flows; thus "fluids" include not only liquids but aeriform substances as well.

led me, in a published memoir,* to name it "igneous fluid" and "the matter of heat." But since then, de Morveau, Berthollet, Fourcroy, and I have determined, in our work together on reforming chemical language, that we ought to eliminate periphrases which tend to impede and lengthen discourse, making it less clear and less precise, and which often even suggest ideas that are not sufficiently correct. For this reason we called the cause of heat, i.e., the exceptionally elastic fluid that produces heat, by the name of *caloric*. This expression not only meets the demands of the system we have adopted, it has an additional advantage. It is compatible with all sorts of opinions since, strictly speaking, we are not even obliged to suppose that caloric is real matter. It suffices, as will be better appreciated after reading what follows, that [6] it should be considered a repulsive cause that separates particles of matter, and so the effects of this cause may be looked at in an abstract, mathematical way.

1.9 Is light a modification of caloric, or is caloric a modification of light? We cannot possibly decide given the present state of our knowledge. This much is certain, though: in our system we have made it a law to admit only facts and to avoid going beyond the facts whenever possible. So in our system, things should be given different names, provisionally, only if they produce different effects. Thus, we will distinguish light from caloric, but we will admit nonetheless that light and caloric have qualities in common and that in some circumstances they combine in almost the same way and produce some of the same effects.

* *Recueil de l'Académie*, 1777, p. 420.

1.8 *igneous*: Fiery; from Latin *ignis*, fire.

we ought to eliminate periphrases: In classical rhetoric, *periphrasis* or *circumlocution* denotes a roundabout way of speaking. Lavoisier and his collaborators argued that many traditional chemical names consisted of numerous words rather than few because the essential makeup of their corresponding substances was unknown; but that it was now both possible and necessary to replace such names by shorter and more meaningful terms.

caloric: The term *calorique* was originated by Lavoisier; from Latin *calor*, heat.

an abstract, mathematical way: Here, the endeavor to reason about the characteristics of an effect without making specific assumptions concerning its cause. Lavoisier will reflect again on such reasoning in paragraph 1.45 below.

1.9 *Is light a modification of caloric?*: That is, is light a *form* of caloric, rather than an essentially different entity? Since heat and light are so often found together, it is reasonable to ask whether one is just a special form of the other.

1.10 The foregoing already would suffice well enough to fix the idea that ought to be attached to the word "caloric." But I must still perform a more difficult task: I must find a way to convey proper ideas of the manner in which caloric acts on bodies. Since this subtle matter penetrates all of the interstices of all substances we know, and since there are no vessels from which it cannot escape, which means that nothing can contain it without some loss, it is only possible to understand its properties through its effects which, for the most part, are fleeting and hard to grasp. In the case of things that can be neither seen nor felt, it is of the utmost importance to guard against flights of the imagination, since the imagination always tends to overleap the truth, and can hardly be confined within the limits circumscribed by facts.

[7]

1.11 We have just seen that the same body becomes a solid, a liquid, or an aeriform fluid according to the quantity of caloric that permeates it; or, to put this more rigorously, according to whether the repulsive force of the caloric is equal to, stronger than, or weaker than the force of attraction among the particles.

1.12 But if only these two forces existed, there would be no range of temperature at which bodies would be liquids, and they would pass abruptly from a solid state to a state of elastic, aeriform fluid. Thus water, for instance, would begin to boil at the moment it ceased to be ice. It would convert into an aeriform fluid, and its particles would disperse throughout space indefinitely. That this does not happen is because a third force, the pressure of the atmosphere, inhibits this dispersion. It is because of this pressure that

[8]

1.10 *to fix the idea*: That is, to determine or specify the idea.

1.12 *these two forces*: That is, the attractive force of particles of the substance for one another, and the repulsive force exerted by caloric as it insinuates itself between those particles.

if only these two forces existed, there would be no ... liquids: Since liquids *flow*, they must contain sufficient caloric to have overcome the mutual attraction of their particles. But if their particles are not attracting one another, what keeps them from dispersing indefinitely throughout space? Yet liquids have a definite volume. Thus without some third force capable of holding the particles together, *liquids* would be impossible.

the pressure of the atmosphere: Torricelli filled a long tube with mercury, then inverted its open end into a basin of mercury; the mercury column fell to a height of about 30 inches, regardless of the length of the tube—and therefore no matter how tall the column had been originally. *Something* must be balancing the weight of the column; he argued that only the pressure of the atmosphere upon the surface of the mercury reservoir could perform that function.

30 in.

water remains in a liquid state between 0 and 80 degrees of the French thermometer. The quantity of caloric that water absorbs at these temperatures is insufficient to overcome the pressure exerted by the atmosphere.

1.13 One can see then that without the pressure of the atmosphere, there would be no stable liquids. We would see bodies in this state only at the very point of melting, for if they were to absorb even the slightest additional heat their parts would separate from each other and disperse. What is more, without the pressure of the atmosphere we could not even properly speak of aeriform fluids. For at the very moment at which the repulsive force of caloric prevailed over the force of attraction, the particles would fly away from each other and disperse indefinitely unless their own weight were to bring them back together so as to form an atmosphere.

1.14 A few simple reflections on well-known experiments should be enough to help us see the truth of what I have just claimed. And this is also confirmed in a straightforward way by the following experiment, which I [9] have already explained in detail to the Academy in 1777.*

1.15 A narrow glass vessel, A, set on its base, P, is filled with sulphuric ether (Plate VII, Figure 17).** This vessel should have a diameter of no more than 12 or 15 lines and should be about 2 inches tall. It should be covered with a moistened bladder which has been fastened to it very tightly by a thick wire wrapped around its neck many times. For greater security, a second bladder should be placed over the first and fastened in the same manner. The vessel should be completely filled with ether so that no air remains between the liquid and the bladder. It

---

* *Recueil de l'Académie*, 1777, p. 426.

** Later I will give a definition for the liquid that is called "ether," and I will explain its properties. I will only say for now that "ether" refers to a highly volatile, inflammable liquid with a specific weight much less than that of water, even less, in fact, than spirit of wine.

1.15 *sulphuric ether*: The liquid produced by distilling alcohol ("spirit of wine") with sulphuric acid. As Lavoisier states in his footnote, it is both volatile and flammable.

12 or 15 lines: the *line* (French *ligne*) is a unit of length equal to .089 inch. See the table of French pre-revolutionary weights and measures in the Appendix.

a moistened bladder: Animal bladders are highly elastic and can be inflated like modern rubber balloons. Pigs' bladders, especially, were used for pretty much the same purposes as today's balloons—both utilitarian and entertaining. Latex balloons were produced in the 1840's but were not widely manufactured until the 1930's.

is then placed beneath the receiver, BCD, of an air pump, the top of which has been covered with a leather case through which passes a rod, EF, with a very sharp blade or point at the bottom, F. A barometer, GH, should be attached to the same receiver.

Fig. 17.

[10]   1.16 Once all of this has been arranged, the chamber is evacuated. Then the pointed rod, EF, is lowered until it punctures the bladder. Instantly, the ether begins to boil away with astonishing rapidity. It evaporates, and is transformed into an elastic, aeriform fluid, which completely fills the receiver. If there is enough ether that some drops of it remain in the vial after the evaporation has stopped, the elastic fluid produced can support, in winter, an 8 to 10 inch column of mercury in the barometer attached to the air pump, or a 20 to 25 inch column in the heat of summer. It is possible, in order to make the experiment more complete, to introduce a little thermometer into the vessel A which contains the ether. The temperature drops considerably while the ether is evaporating.

1.15, cont. *the receiver, BCD, of an air-pump*: An odd name, one might think, for this vessel since, so far from "receiving" anything, it will have air pumped *out of* it. But containers of this size and style had long been used for collecting the products of distillation, and so had acquired the name "receiver" with good reason. Although the utensil later found other uses, it was still crafted in the same way and was therefore, as far as its makers were concerned, still a "receiver." In 1660 Robert Boyle described a container "which we, with the Glass-men, shall often call a Receiver, for its affinity to the large Vessels of that name, used by Chymists" (Oxford English Dictionary).

leather case: This forms an airtight seal about rod EF, while allowing it to move up and down.

a barometer, GH: A mercury-filled tube, sealed at the top as shown here (compare the sketch accompanying the comment to 1.12).

1.16 *the elastic fluid ... can support ... an 8 to 10 inch column of mercury*: When the apparatus is cold (in winter), the aeriform ether exerts sufficient pressure within the receiver to support the weight of 8–10 inches of mercury. But why is the pressure so much greater in summer, as he goes on to report? Since accumulation of caloric is the cause of the sensation of heat (1.8), a higher temperature indicates the presence of a greater quantity of caloric. And since caloric is itself preeminently elastic, a greater quantity confined in the same space will exert greater pressure. An accumulation of caloric will also cause more liquid to evaporate, as Lavoisier outlined in paragraph 1.11.

1.17 In this experiment the weight of the atmosphere, which ordinarily presses on the surface of the ether, is eliminated, and that is all. The effects that follow from this clearly prove two things. First, at the temperatures in which we live, ether would always remain in the state of an aeriform fluid if it were not for the pressure of the atmosphere. Second, the passage from the liquid state to the aeriform state is accompanied by considerable cooling due to the fact that some of the caloric, either free or at least in equilibrium with the surrounding bodies, combines with the ether during vaporization in order to convert it into an aeriform fluid.

[11]

1.18 This same experiment succeeds with all evaporable fluids, such as spirit of wine (i.e., alcohol), water, or even mercury, but with this difference: the atmosphere of alcohol that forms under the receiver supports, in winter, a column of mercury of only one inch above the level of the barometer attached to the air pump, and four or five inches in summer. Water can only support a column of a few lines, mercury a fraction of a line. Less fluid evaporates when working with alcohol than when working with ether, even less in the case of water, less still in the case of mercury. Because of this, less caloric is involved and so there is less cooling, and this agrees perfectly with the results of experiments.

---

the temperature drops considerably while the ether is evaporating: A liquid evaporates (becomes aeriform) by acquiring additional caloric (1.11). This caloric can only be supplied by the surrounding bodies—especially by the remaining liquid. As that residue loses caloric, its temperature drops.

1.17 *the weight of the atmosphere ... is eliminated*: Before the receiver was evacuated, the surface of the liquid ether in vessel A was pressed by the full weight of the atmosphere; and that pressure was maintained by the bladders tied over A, even after the air was pumped out of the receiver. But the instant the bladders were punctured, the liquid surface became exposed to an environment from which the atmosphere's weight had been eliminated. As Lavoisier points out, elimination of atmospheric pressure is all that was necessary to allow the liquid to become aeriform.

caloric ... combines with the ether: Notice that all of the phenomena discussed so far appear to involve caloric that is somehow attached to, or combined with, other substances. Caloric can, evidently, pass from one substance to another—as when a hot body warms a cold one by contact—but we have not seen any indications that caloric can accumulate *on its own*, apart from another material. Why would caloric always have to be united with some other substance? And what might be the nature of such combination? Lavoisier will say more about the relations between caloric and other materials in paragraph 1.36.

1.18 *this same experiment succeeds with all evaporable fluids*: But he finds that each liquid develops its own characteristic pressure in the aeriform state. When they are at the same temperature, *ether, alcohol, water*, and *mercury* exhibit aeriform pressure in descending order.

1.19 Another kind of experiment proves just as clearly that the aeriform [12] state is a modification of bodies and that it depends on the temperature and pressure to which they are exposed.

1.20 In a memoir that we read to the Academy in 1777 but have yet to publish, Laplace and I showed that when ether was submitted to a pressure of 28 inches of mercury, that is to say, to a pressure equal to that of the atmosphere, it began to boil at 32 or 33 degrees of a mercury thermometer. De Luc has made analogous experiments with spirit of wine and has established that it begins to boil at 67 degrees. And everyone knows that water begins to boil at 80 degrees. Since boiling is nothing but the evaporation of a fluid, or the point at which it passes from a liquid to an elastic, aeriform state, it was clear that when ether is kept at a temperature higher than 33 degrees and at an ordinary degree of atmospheric pressure, it ought to exist in the state of an aeriform fluid, and that the same thing ought to happen for spirit of wine above 67 degrees and water above 80. All of this is completely confirmed by the following experiments.*

Fig. 15.

[13]   1.21 I filled a large vessel, ABCD, with water at a temperature of 35 or 36 degrees (see Plate VII, Figure 15). In order to more easily appreciate what is happening inside, I assume that the flask is transparent. It is possible to immerse one's hands for a long time without discomfort in water at this temperature. I submerged bottles F and G upside down in the jar, first filling them and then upending them so that they stood on their necks at the bottom.

1.22 This being done, I filled a very small long-necked flask, the neck *abc* of which had been bent twice, with suphuric ether. I submerged this flask in vessel ABCD, and managed to introduce the neck of the flask under the neck of one of the bottles, F, in the manner illustrated in the figure. As soon as it began to feel the effects of the heat, the ether began to boil, and the caloric that combined with it converted it into an elastic, aeriform fluid, with which I filled several bottles like F and G.

* *Mémoires de l'Académie*, 1780, p. 335.

1.19 *the aeriform state is a modification of bodies*: In becoming aeriform, a body merely takes on a different form; it does not change its essential nature or become a different substance. Recall Lavoisier's similar use of "modification" in paragraph 1.9.

1.20 *28 inches of mercury ... equal to the pressure of the atmosphere*: Lavoisier's "inch" is actually the *pouce*, equal to 1.066 modern inches (see the table of French prerevolutionary weights and measures in the Appendix). So his determination of atmospheric pressure would actually be 29.8, expressed in modern inches.

1.23 Here is not the place to examine the nature and properties of this aeriform fluid, which is extremely flammable; but without relying prematurely on knowledge I ought not to assume the reader possesses, and focusing only [14] on the matter before us at the moment, I will observe, on the basis of this experiment, that it is almost the case that ether can only exist in an aeriform state on the planet we inhabit. If the pressure of our atmosphere were equivalent to a column of 20 or 24 inches of mercury rather than 28, we could not obtain ether in a liquid state, at least during the summer. Accordingly, it would be impossible to find liquid ether on mountains at higher elevations, for it would convert to gas as soon as it formed unless very strong flasks were used to condense it and it was both pressurized and cooled. Finally, since the temperature of blood is just about the temperature at which ether passes from a liquid to an aeriform state, ether ought to evaporate as soon as it is introduced into the body; and it is quite likely that the medicinal properties of ether depend on this, so to speak, mechanical effect.

1.24 These experiments succeed even better with nitrous ether, because it evaporates at a lower temperature than sulphuric ether. As for alcohol or spirit of wine, the experiment required to obtain it in an aeriform state is somewhat more difficult, because this fluid only evaporates at 67 degrees Réaumur. So it [15] is necessary to submerge the flask of alcohol in water that is almost boiling, at which temperature the water is too hot to place one's hands in it.

1.25 It was clear that the same ought to hold for water, that this fluid should also convert into gas when exposed to greater heat than is required to make it boil. But although we were convinced that this was true, Laplace and I thought we ought to confirm it with a direct experiment, the result of which was as follows. We filled a closed tube of glass, A, with mercury, turned it

---

1.22 *sulphuric ether*: The same liquid used in the previous experiment (1.15).

1.23 *the medicinal properties of ether*: If Lavoisier means its effects as a general anesthetic, his conjecture is probably erroneous. The physiological processes associated with anesthesia are only imperfectly understood; but they are unlikely to be "mechanical," as Lavoisier suggests, in any significant degree.

1.24 *nitrous ether*: It is prepared in the same way as sulphuric ether, but using nitric instead of sulphuric acid. Lavoisier will discuss both these acids in Chapter Six.

degrees Réaumur: Degrees of "the French thermometer" (1.5) originally introduced by René Antoine Ferchault de Réaumur in 1730.

1.25 *gas*: Lavoisier previously (1.5) offered *gases* and *vapors* as examples of aeriform fluids. Some chemists regarded gases and vapors as different in kind, but for Lavoisier there is no distinction. He will adopt "gas" as the general term in paragraph 1.29.

upside down, and placed it in a mercury-filled dish, B, which we slid beneath it (see Plate VII, Figure 5). Into the tube we introduced about 2 gros of water, which stood at height CD and sat above the surface of the mercury. Then we submerged all of this in a large iron boiler, EFGH, which had been placed on top of a furnace, GHIK. This boiler was full of boiling brine, the temperature of which was above 85 degrees. As we know, water rich in salts can be raised to a greater degree of heat than boiling water.

Fig. 5.

[16]

1.26 As soon as the 2 gros of water placed in the upper part, CD, of the tube reached a temperature of about 80 degrees it began to boil, and instead of occupying the small space ACD, it was converted into an aeriform fluid that filled the tube completely. The mercury even descended a little below its ordinary level, and the tube would have flipped over if it had not been so thick and, for this reason, very heavy, and if it were not also attached to the dish by iron wire. Moments after we removed the tube from its salt-water bath, the water condensed and the mercury rose again. But the water resumed its aeriform state as soon as the assembly was submerged again.

1.27 And so there are quite a few substances which convert into aeriform fluids at temperatures very close to those in which we live. We will see shortly that there are others such as marine or muriatic acid, volatile alkali (ammoniac), carbonic acid (fixed air), sulphurous acid, and so on, which always

---

1.25, cont. *2 gros of water*: About $7\frac{1}{2}$ grams or $\frac{1}{4}$ US ounce. See the table of French pre-revolutionary weights and measures in the Appendix.

water ... stood at height CD: A bent pipe like that used in (1.22) would serve to bring water into tube A from the bottom. The water would rise to the top at CD, displacing some of the mercury that formerly filled the tube.

salts: Originally, "salts" denoted substances that resembled common salt in appearance or in properties such as solubility. In Chapters Sixteen and Seventeen Lavoisier will distinguish the particular class of "neutral" salts.

raised to a greater degree of heat than boiling water: A boiling liquid does not increase in temperature no matter how intensely it may be heated; it merely boils more furiously and vaporizes at a greater rate. That stability is the reason why the boiling remain in an aeriform state under the ordinary conditions of heat and pressure of the atmosphere.

1.28 All of these facts, of which I could easily give many additional [17] examples, entitle me to assert as a general principle what I have already said above: that almost all bodies in nature can exist in three different states, in a solid state, a liquid state, and an aeriform state, and that these three states of the same body depend on the amount of caloric with which it is combined. From now on, I will designate aeriform fluids by the general term "gas," and so for each kind of gas I will say that caloric, which assumes in some way the role of solvent in every case, should be distinguished from the substance with which it combines and which forms the gas's base.

1.29 We have been obliged to give names to these bases of gases, bases which are still not widely known. I will point them out in Chapter Four of

---

point of water is a useful standard for thermometry. But water containing dissolved substances generally boils at a higher temperature than does pure water; Lavoisier's brine (sea-water) boiled at 85° Réaumur; pure water would have boiled at 80°.

1.28 *the role of solvent*: When one substance dissolves in another, the ingredient that dissolves is the *solute*; the substance in which it dissolves is the *solvent*. When, for example, salt dissolves in water, salt is the solute, water is the solvent.

the substance with which it combines and which forms the gas's base: The *base* of a gas is that solid or liquid substance which, by combining with sufficient caloric, takes on the gaseous form. Water, for example, is the base of steam. In Lavoisier's image of the gaseous state as a *solution*, if caloric is the solvent, the base is the solute.

Lavoisier generally uses the term "base" to mean something like *the foundation upon which a substance is formed*. The "base" of a substance is, at least generally, the base *of that substance only*—or at most a few others.

Note: Readers who have had any exposure to chemistry, or even to gardening, will have become familiar with a modern sense of *base* as the opposite of *acid*. Although the modern usage derives to some extent from Lavoisier's, his notion of a "base" is far more general. For Lavoisier, *every acid begins with its base*, which is thus a constituent of the acid rather than a counterpart to it.

Notice also that Lavoisier states that the base of a gas *combines* with caloric, even though he is comparing gases to *solutions*. Today we distinguish sharply between *solutions*—in which the same quantity of solvent can take up a continuously varying amount of solute—and *true compounds*, which are characterized by *fixed proportions* of the combining substances. Though Lavoisier is exquisitely aware that many substances combine in definite ratios, he has no way, as yet, to discover *how* substances associate with one another; so for him it is far from obvious that the distinction between definite ratios and varying ratios of combination is a fundamental one. He will, in fact, pay attention almost exclusively to fixed-ratio combinations, but probably not for any strict theoretical reasons.

this work after I have given an account of some of the phenomena associated with heating and cooling bodies and have supplied more precise ideas about the constitution of our atmosphere.

[18]

1.30 We have seen that the particles of all bodies in nature are in a state of equilibrium between attraction, which tends to bring them together and unite them, and the influence of caloric, which tends to separate them. This implies that caloric not only surrounds every part of every body, it also fills the interstices between their particles.

1.31 It is possible to form an idea of these tendencies if one imagines a vessel filled with little balls of lead into which a very fine powder, like sand, has been poured. In this case, it is clear that the sand will spread out uniformly and fill the spaces left between the balls. The balls in this example are to the sand as the particles of bodies are to caloric, except that the balls in the example touch each other while the particles do not, but rather maintain a little distance from one another due to the influence the caloric exerts.

1.32 If one were to substitute hexahedrons or octahedrons or some other regular bodies of equal solidity for the round balls, there would no longer be the same amount of space between the objects in the jar and it would no longer be possible to fill that space with the same amount of sand. It is the same for all bodies in nature. The space between the particles of different bodies is not of equal volume, and the volume depends on the shape of the particles, their size, and on the distance from each other at which they are maintained according to the ratio between their force of attraction and the repulsive force exerted by the caloric.

[19]

1.33 It is in this sense that the phrase "the capacity of bodies for containing the matter of heat" should be understood, which is a particularly apt expression introduced by the English natural philosophers who were the first to hold precise notions about this subject. An example of what happens in the case of water, and several reflections on the manner in which water saturates and penetrates bodies, will make all of this more intelligible. Clearer understanding of abstract things is always possible with the aid of sensible comparisons.

1.32 *hexahedrons or octahedrons*: Here, six- or eight-sided solid bodies.

no longer be the same amount of space: Since bodies having flat sides would not pack together the same as round balls, the amount of space between them would also not be the same as the space between round balls.

1.33 *the capacity of bodies for containing ... heat*: Joseph Black found that when equal weights of hot mercury and cool water were stirred together, the mixture attained a common temperature much closer to the initial temperature of the water than to that of the mercury. But the heat lost by the mercury must have been gained by the

1.34 If you submerge pieces of different kinds of wood, equal in volume (say a cubic foot) in water, the fluid will seep into the pores bit by bit. The pieces of wood will swell and gain weight, but each piece will admit a different amount of water into its pores. The lightest and most porous pieces will take in more, the heavier and denser pieces only a very little. Finally, the proportion of water that they soak up will also depend on the nature of the particles that constitute the wood—on the greater or lesser affinity that they have for water; and very resinous wood, for example, despite being very porous admits very little. So one could say that different kinds of wood have different capacities for receiving water, and it is even possible to know how much water a piece of wood has absorbed by considering how much weight it has gained. Since, however, we do not know how much water it contained before it was immersed, it will be impossible to know the absolute quantity it contains in the end. [20]

1.35 The same considerations apply to bodies immersed in caloric, though it should be observed that water is an incompressible fluid, while caloric is endowed with great elasticity, that is to say, in other words, that the particles of caloric have a strong tendency to separate from each other when some force has drawn them together. It is clear that this factor must lead to very notable differences in the results.

---

water; so equal quantities of heat gained or lost resulted a *small* temperature change for water, but a *large* temperature change for mercury.

Now if you pour a few ounces of water into a drinking glass, the water level might rise by several inches. But the same amount of water poured into a capacious mixing bowl will produce a much smaller rise. Similarly, Black interpreted a body's change in temperature when it gained or lost heat as an indication of that body's *capacity for heat*: the smaller the temperature change (for a given quantity of heat), the greater its capacity.

the matter of heat: This expression was widely employed to indicate the treatment of heat as a substance. Lavoisier employed it in a 1777 memoir but, for the reasons he noted in paragraph 1.8, subsequently adopted the term "caloric."

the English natural philosophers: Prominent among these in the eighteenth century would be Joseph Black, Joseph Priestley, and John Dalton. Among them, however, it is probably Black who may best be said to have held "precise notions" concerning the heat capacity of bodies.

1.35 the *particles of caloric have a strong tendency to separate from one another when some force draws them together*: That is, the particles of caloric *resist* any effort to squeeze them together. This is caloric's "great elasticity," just as a foam rubber ball resists the effort to compress it and regains its original volume as soon as the compression is removed. Lavoisier will make this tendency to separate a *definition* of elasticity in paragraph 1.45 below.

1.36 Now that we have brought these issues to such a point of clarity and simplicity, it will be easy for me to expound the ideas that ought to be attached to the expressions "free caloric," "combined caloric," the "specific quantity of caloric" contained in different bodies, "capacity to contain caloric," "latent heat," and "sensible heat." These expressions are by no means synonymous: each has a strict and determinate sense, in light of what I have just explained. So I shall attempt to fix the senses of these terms with the following definitions.

[21]

1.37 "Free caloric" is caloric which is not involved in any combination with any other body. But since we live amidst a system of bodies to which caloric adheres, it so happens that we can never obtain this principle in an absolutely free state.

1.38 "Combined caloric" is caloric which is locked up within bodies by the force of affinity or attraction, and which constitutes an aspect of the substance, and even the solidity of bodies.

1.39 By the expression "specific caloric" of bodies is meant the quantity of caloric required to raise the temperature of bodies, equal in weight, the same number of degrees. This quantity of caloric depends upon the distance between the particles of bodies—upon their lesser or greater adherence; and it is this distance, or rather the space that results from it, which (as I have already noted) has been named "the capacity to contain caloric."

[22]

1.40 "Heat," considered as a sensation, or, in other words, "sensible

---

1.37 *this principle*: That is, *caloric*. For Lavoisier, a "principle" is primarily a *rule* or *standard*—as, for example, the "principle" that *heated bodies expand*, which he cited in paragraph 1.1. But he also uses the term to denote a substance that, so far as we can tell, is *simple* and not a composition of other substances. Such is its usage here. Recall that in his Preliminary Discourse, paragraph 0.20, he treated "principles" and "elements" as virtually synonymous.

1.39 *the space that results ... has been named "the capacity to contain caloric"*: Imagine equal volumes of two solids or liquids A and B; and suppose A has more space between its particles than does B. Then equal amounts of caloric, added to both, will be *more highly compressed* in B and must then exert greater repulsive force in B than in A. But Lavoisier has implicitly associated a body's expansion, and hence its temperature, with this repulsion (1.2, 1.7, 1.12). Adding equal amounts of caloric to both bodies, then, would produce a greater temperature change in B than in A. Black (1.33, comment) would therefore have declared A to have *greater capacity for heat* than B; and Lavoisier is pointing out that the substance that has greater capacity for heat is also the substance that has *more space between its particles*.

heat," is nothing else but the effect produced on our organs of sense by the caloric emanating from surrounding bodies.

1.41 In general, we only experience sensations when there is some movement. We can put this down as an axiom: "no movement, no sensation." This general principle naturally applies to the sensation of cold and heat: when we touch a cold body, the caloric, which seeks equilibrium in all bodies, passes from our hand into the body we are touching, and we feel the sensation of cold. The opposite effect occurs when we touch a hot body: caloric passes from the body into our hand and we experience a sensation of heat. If the body and our hand are the same or approximately the same temperature, we feel nothing, neither cold nor heat, for in this case there is no movement, no caloric is transmitted, and so once again there is no sensation without a movement to cause it.

1.42 When the thermometer rises, it is a certain indication that free caloric is spreading into the surrounding bodies. The thermometer is just one of these bodies, and it receives its portion of caloric in accordance with its mass, and in accordance with its capacity for containing caloric. The thermometer only [23] registers a change in the system of bodies to which it belongs, a displacement of caloric. It indicates nothing else but the amount of caloric it has received and does not measure the total quantity disengaged, displaced, or absorbed.

1.43 Laplace invented the simplest and most precise way to measure the quantity of caloric disengaged. This method is described in a memoir,* and a summarized explanation of the procedure appears at the end of this work.

* *Mémoires de l'Académie*, 1780, p. 364

1.42 *The thermometer ... indicates nothing else but the amount of caloric it has received*: Even though we commonly speak of the thermometer as measuring the temperature of a body to which it is applied, or of a liquid in which it is immersed, it properly should be said to indicate *its own* temperature—for the length of its mercury column expands and contracts according to the amount of caloric it has received from the body or liquid being measured.

1.43 *a summarized explanation of the procedure appears at the end of this work*: The device used to measure the quantity of caloric that different substances release is the *ice calorimeter*. Lavoisier's summary appears in Part III of the *Treatise*; briefly, the procedure is this: Ice fills the space *bb* between the outer compartment and the inner chamber A. The body under test is placed in A, and the apparatus covered with an insulating lid. The test body communicates caloric to the ice, melting it; and the amount of water that drips from spout *y* is a measure of the amount of caloric released.

It involves placing a body or several bodies releasing caloric in the middle of a hollow sphere of ice. The quantity of ice melted is an exact expression of the amount of caloric released. It is then possible, with the help of equipment that we have constructed for the purpose, to determine on the basis of the number of degrees indicated on the thermometer, not (as has been purported) the bodies' capacities for containing caloric, but the relationship between increases and decreases in their capacities for caloric in accordance with particular degrees of temperature. It is easy, with the same equipment, and by many different combinations of experiments, to determine the quan-

[24] tity of caloric necessary to convert solid bodies to liquid and liquid bodies to aeriform fluids; and, *vice versa*, that amount of caloric which the elastic fluids lose when they again become liquids, and these, when they again become solids. Someday, once there have been a sufficient number of experiments, we may be able to determine the ratio of caloric in each type of gas. I will devote a special chapter to an account of the principal results we have obtained in this area so far.

1.44 Before concluding this chapter, it only remains for me to say a word about the cause of the elasticity of gases and vaporous fluids. It is not hard to see that this elasticity is related to the elasticity of caloric, which seems to be the most elastic body in nature. Nothing is simpler than to imagine that a body becomes elastic through its combination with another body which is itself endowed with that property. But it must be acknowledged that this would be to explain elasticity by elasticity, that this merely shifts the problem back a step, and that it remains, as before, to explain what elasticity truly is, and why caloric is elastic. Considered in the abstract, elasticity is simply the prop-

[25] erty that bodies have of separating from one another when they are forced together. This tendency that particles of caloric have—that of separating themselves—extends even to very great distances. To be convinced of this, consider the fact that the air is susceptible to very high compression, and this means that particles of air are already very distant from one another since the distance between them must be at least sufficient for them to get so much closer to one another. Now these particles of air, which are already quite distant from each other, tend to separate even further: indeed, if Boyle's vacuum

---

1.44 *elasticity is simply the property...*: Lavoisier noted the tendency of particles to separate, and so resist being pressed together, in paragraph 1.36 above. Now he identifies that property as the *definition* of elasticity.

Boyle's vacuum: Robert Boyle's improved air-pump allowed him to evacuate the air from a large "receiver," and by such means to demonstrate that air was necessary for is created in a large receiver, the last portion of air that remains spreads out uniformly throughout the whole volume of the vessel. No matter how large the vessel may be, the air fills it entirely and presses against its sides. Now this effect can only be explained by assuming that the particles endeavor to spread out in all directions, and it is unknown whether there is a distance beyond which this phenomenon no longer occurs.

1.45 There is, then, a real repulsion between the particles of elastic fluids, or at least everything occurs in the same way it would if there were such a repulsion. It seems permissible, then, to conclude that the particles of caloric repel one another. Once we admit this force of repulsion, explaining the formation of aeriform fluids or gases becomes very simple. But at the same time it must be admitted that a repulsive force which acts between tiny particles at great distances from one another is hard to conceive.

1.46 It may seem, perhaps, more natural to suppose that the particles of caloric attract one another more than the particles of bodies do, and that the

[26]

---

the transmission of sound, for the combustion of a candle, and for the maintenance of life.

1.45 *at least everything occurs in the same way it would if...*: Since there is no way to observe the particles of elastic fluids directly, Lavoisier cannot positively affirm that they exert repulsive forces on one another. Their mutual repulsion must remain a hypothesis; but since (as Lavoisier has shown) it would explain so many things if it were true, the hypothesis is a very useful one.

To reason freely on the basis of this hypothesis—drawing conclusions from it without necessarily claiming certainty for it—is an example of treating phenomena "in an abstract, mathematical way," as Lavoisier expressed it in paragraph 1.8. Evidently, for him, a *mathematical* way need not involve symbols or equations. What more inclusive sense of "mathematical" might he have in mind?

a repulsive force which acts ... at great distances ... is hard to conceive: One might regard *any* force acting at great distances, whether repulsive or attractive, as "hard to conceive"; Newton had to face such an objection to his theory of universal gravitational attraction. Is there any reason why a *repulsive* force would be particularly challenging?

1.46 *more natural to suppose that the particles of caloric attract one another more than the particles of bodies do*: On this alternative there is no need to hypothesize *repulsion* at all: if *all* particles—those of caloric as well as those of bodies—*attract* rather than repel one another, then in their competition to congregate, whichever particles have less mutual attraction will be driven apart as those with stronger attraction succeed in coming together.

But why would anyone find this alternative supposition "more natural"? Some possible reasons: (i) the alternative hypothesis is simpler because [OVER] particles of bodies only separate from one another because of the attractive force that causes particles of caloric to unite. Something analogous to this phenomenon happens when a dry sponge is submerged in water: it swells, and its particles separate from one another as water fills the space between them. It is clear that the sponge, in swelling, has acquired a greater capacity to contain water than it previously had. But should we say, in this case, that the water communicates a repulsive force to the particles of the sponge that tends to separate them from one another? Certainly not: in this case only attractive forces are at work. These forces are, first of all, the weight of the water and the effort that it exerts in all directions like any fluid; second, [27] the attractive force between the particles of water; third, the attractive force between the particles of the sponge, and finally the reciprocal attraction between the particles of the water and the sponge. It is easy to understand that the explanation of the phenomenon depends upon the intensity and relation of all these forces. It is likely that the separation of the particles of bodies by caloric depends in the same way on some combination of different attractive forces. It is the result of these forces that we attempt to express, in a way that is more concise and better conforms to the imperfect state of our understanding, by saying that caloric communicates a repulsive force to the particles of bodies.

---

it posits only one force (attraction) rather than two (attraction and repulsion); (ii) Newton's gravitation theory has already given us good reason to suppose that all particles attract one another; (iii) Lavoisier's example of the sponge, to which he devotes the rest of the paragraph, helps make the alternative hypothesis visualizable, and so escapes the objection "hard to conceive" in paragraph 1.45.

Notice how detailed the foregoing speculation has become! Since in fact we have no access to the individual particles of bodies, is such meticulous conjecture really compatible with the "abstract, mathematical way" (1.8; 1.45, comment)? Lavoisier will end this paragraph with what appears to be a corrective effort.

express in a way that ... better conforms to the imperfect state of our understanding: Lavoisier implies a profound maxim here. It is better to characterize things in a general and imprecise way—if such indeed is the state of our knowledge about them—than to weave elaborate and detailed hypotheses that misleadingly imply a level of precision that our thinking does not actually possess. Some eighty years after the publication of Lavoisier's *Traité*, James Clerk Maxwell in his own *Treatise* would follow an equivalent maxim when propounding a new mathematical science of electricity and magnetism.

caloric communicates a repulsive force to the particles of bodies: Lavoisier reasserts his original supposition, which represents the phenomena in a conveniently concise way, and without *explaining too much*.

Chapter Two

Overall Views on the Formation and Constitution
of the Earth's Atmosphere

2.1 The account of the formation of aeriform elastic fluids, or gases, that I have just presented sheds great light on the way in which the atmospheres of planets, in particular the earth's, were formed in the beginning. It is clear that our atmosphere must be a product and an amalgam of (1) all evaporable substances or, better, all substances which remain in an aeriform state at the temperatures in which we live, and at a pressure equal to a column of mercury 28 inches high; and of (2) all fluids or solids that can be dissolved in this assortment of different gases.

2.2 To better fix our ideas concerning a matter which no one has yet considered carefully enough, let us imagine for a moment what would happen to the various substances which compose the earth if the temperature were to change abruptly. Let us imagine, for example, that the earth were suddenly transported to a much hotter region of the solar system, perhaps somewhere near Mercury, where the normal temperature is probably much higher than that of boiling water. Before long, water, all fluids which evaporate at any temperatures close to that of boiling water, and mercury itself, would begin to expand. They would all convert into aeriform fluids or gases, and would become part of the atmosphere. These new types of air would mix with those already present, resulting in reciprocal decompositions and new compounds, until the principles composing these different airs or gases should achieve a state of equilibrium determined by their respective affinities for one another.

2.2 *new types of air*: Although "air" most often refers to the gaseous substance that surrounds the earth and which animals breathe, it also has a more general meaning, as here, denoting any gas or vapor.

2.3 But we should not lose sight of the fact that this vaporization would be limited. For as the quantity of elastic fluid increased, the weight of the atmosphere would increase proportionally. And since any pressure would pose an obstacle to vaporization, and since the most evaporable fluids can resist evaporating under great heat so long as that heat is counteracted by a proportionally greater pressure, and, finally, since all liquids, even water itself, can sustain a red heat in Papin's machine, it is clear that the new atmosphere would reach such a degree of heaviness that any water which had not yet evaporated would cease boiling and would remain in a liquid state. As a result, even in the case we are imagining, or in any other case of the same kind, the weight of the atmosphere would be limited and could not surpass a certain boundary. We could carry these reflections much further and examine what would happen to stones, to salts, and to many other substances which can be melted and which compose the earth. We could imagine how they would soften, begin melting, and become fluids. But these are digressions, and I hasten to return to the point.

[30]

2.4 If the earth were suddenly placed, instead, in some very cold region, just the opposite would occur. The water that makes up our rivers and oceans, and probably most known liquids, would be converted into solid mountains, and into extremely hard rocks. These would be translucent, homogeneous, and white at first, just like rock crystal. But over time, by mixing with substances of different natures, they would become diversely colored opaque stones.

2.3 *the weight of the atmosphere would increase*: The substances which form earth's atmosphere have weight, even though they are in an aeriform state. As more and more materials take on a gaseous condition, so much greater will be the total weight of those gases composing the atmosphere.

any pressure would pose an obstacle to vaporization: Recall Lavoisier's argument that the pressure of the atmosphere is what prevents immediate vaporization of solids and so makes the liquid state possible (1.13).

water ... can sustain a red heat in Papin's machine: Also known as Papin's digester, pictured here, this early pressure cooker was capable of developing pressures so high that water remained liquid even when the apparatus was heated to a red glow.

2.5 Under this supposition the air, or at least some of the aeriform sub- [31]
stances composing it, would undoubtedly cease to exist as elastic vapors, due
to an insufficient degree of heat. They would therefore return to a liquid state,
and we have no idea of the new liquids they would form.

2.6 These two radical suppositions make it clear that (1) *solidity*, *liquidity*,
and *elasticity* are different states of the same matter: three specific modifications through which nearly all substances can pass in succession, and which
depend only on the degree of heat to which they are exposed, that is, on the
quantity of caloric which penetrates them; that (2) it is very likely that air is
naturally a vaporous fluid, or better put, that our atmosphere is a composition
of all fluids that can exist in a vaporous state of constant elasticity, at what is,
for us, ordinary heat and pressure; and that (3) it would consequently not
be impossible to come across otherwise extremely dense substances in our
atmosphere, even metals, such as, for example, a metallic substance just a bit
more volatile than mercury.

2.7 Among known fluids, some, such as water and alcohol (i.e., spirit of [32]
wine), can mix with each other in any proportion. Others, on the contrary,
such as mercury, water and oil, can only combine with one another momen-
tarily, and they separate when mixed and arrange themselves according to
their specific gravities. The same thing must, or at least can, occur in the

---

2.6 *it would ... not be impossible to come across otherwise extremely dense substances
in our atmosphere*: The density of a substance is measured by the ratio of its weight
to the volume it occupies. All the constituents of the atmosphere are in the gaseous
state, and must therefore have very low densities in comparison to solids and liquids.
But it is possible that some of those gaseous components, if they could be condensed
into liquid form, would prove to be metals like mercury, which is more than 13 times
as dense as water. Lavoisier will examine the constitution of atmospheric air in the
next chapter.

a metallic substance just a bit more volatile than mercury: Mercury exists naturally
in the liquid state; but if some metal were only slightly more volatile than mercury,
the atmosphere would be likely to contain significant quantities of that metal's vapor.

2.7 *arrange themselves according to their specific gravities*: The specific gravity of a
substance is the ratio of its density to the density of water. In an assortment of fluids
that do not mix—or even a solid placed in a fluid—the denser material will displace
the less dense as it descends. Thus water, with a specific gravity of 1, forms a layer
upon the surface of mercury, whose specific gravity is 13.5; Lavoisier observed this
at CD in Plate VII, figure 5 (page 26 above). For the same reason olive oil, with a
specific gravity of .93, will form a layer upon the surface of water. But such layers
would not form if the liquids could mix with one another.

atmosphere. It is possible, even probable, that gases that mix with atmospheric air only with difficulty and tend to separate from it were formed in the very beginning and form constantly since. If these gases are lighter than air, they must collect in higher regions and form layers which float on atmospheric air. The phenomena associated with fiery meteors lead me to believe that there is a layer of flammable fluid like this above the atmosphere, and that the aurora borealis and other fiery atmospheric phenomena take place at the point of contact between these two layers of air. I intend to develop my ideas on this matter in a separate memoir.

2.7, cont. *gases that mix with atmospheric air only with difficulty*: If the particles of gases are as distant from one another as Lavoisier has proposed (1.44), it is hard to understand what sort of "difficulty" could hinder any gas from mixing with any other gas. Nevertheless, such phenomena are observed. Propane gas, widely used for cooking and heating, is denser than atmospheric air under identical conditions. While propane does mix with air, such mixing takes time, especially indoors where there is no wind. To whatever extent the gas has not diffused, it tends to collect in basements, presenting an explosion hazard. Natural gas (largely methane) is less dense than air and so tends to be displaced upwards and diffuse more readily.

aurora borealis: Northern lights. Their underlying causes remain mysterious even today, but they certainly involve some sort of electrical excitation rather than the fiery combustion Lavoisier suggests.

Chapter Three

The Analysis of Atmospheric Air: Its Resolution into Two Elastic Fluids, One Respirable and the Other Non-respirable

3.1 This much can be said about the constitution of the atmosphere *a priori*: it should be made up of a collection of all gases which remain in an aeriform state at what is, for us, ordinary temperature and pressure. These fluids form a nearly homogeneous mass from the surface of the earth to the greatest height that has been reached, and the density of this mass decreases in inverse proportion to the weight it bears. But as I have already said, it is possible that this first layer of the atmosphere is covered with another very different fluid, or several others.

3.2 It now remains for us to determine the number and the nature of the elastic fluids which compose this lowest layer of the atmosphere, the layer we inhabit. Here experiment will light the way. Modern chemistry has made great strides in this context, and the points I will relate will make it clear that atmospheric air is perhaps the most precisely and rigorously analyzed substance of its class.

3.3 In general, chemistry furnishes two means for determining the nature of the constituent parts of bodies: composition and decomposition. When, for example, water and spirit of wine (i.e., alcohol) are combined, creating as a result the liquid known commercially as eau-de-vie, it should be inferred

---

3.1 *the density ... should decrease in inverse proportion to the weight it bears*: Expressed more straightforwardly, the density of this collection of gases should *increase* in *direct* proportion to the weight that bears on it.

Lavoisier is drawing on the law of compressibility of gases that had been discovered by Boyle in England, and in France by Marriotte: *when the temperature of a gas is kept constant, the volume of the gas is inversely proportional to the pressure to which it is subjected*. Then since the density of a given weight of gas is inversely as its volume (2.6, comment), and the pressure is likewise inversely as the volume, the density and the pressure should be proportional to one another.

Lavoisier is also aware of a relation between a gas's volume and its *temperature*, a relation that would later be known as Charles' Law. He will employ that relation in the measurements mentioned in paragraph 3.7.

that eau-de-vie is composed of alcohol and water. But the same conclusion can be reached through decomposition, and in chemistry, complete satisfaction requires, in general, the combination of both of these types of proof.

3.4 In the analysis of atmospheric air, we have the advantage of being able both to decompose and to recombine it. I will limit myself here to reporting the most conclusive experiments that have been made in this regard. Almost all of them are my own, either because I was the first to make them, or because I repeated them with the new aim of analyzing atmospheric air.

[35]   3.5 I took a flask, A (Plate II, Figure 14), about 36 cubic inches in volume, with a very long neck, BCDE, about 6 or 7 lines in interior diameter. I bent it in the manner illustrated (Plate IV, Figure 2) so that when placed in

the furnace, MMNN, the top of its neck, E, could remain inserted beneath a bell jar, FG, which had been placed in a bath of mercury, RRSS. I introduced into this flask four ounces of very pure mercury, and then, by sucking on a siphon that I had inserted beneath the bell jar, FG, I raised the mercury to the level LL. I glued a band of paper to the bell jar in order to mark this height carefully and took precise observations of the barometer and thermometer.

3.6 With this equipment in place, I lit a fire in the furnace MMNN, and I tended it almost continuously for 12 days, keeping it just below the boiling temperature of mercury.

3.7 Nothing remarkable occurred on the first day. The mercury, though it did not boil, constantly evaporated. It covered the interior of the vessels with droplets, which were at first very fine, but which gradually enlarged. Once [36]   they had acquired a certain volume, they fell back again and rejoined the rest of the mercury at the bottom of the vessel. On the second day, I began to see little red particles swimming on the surface of the mercury, which

---

3.5 *the level LL*: In Figure 2, the letters LL actually mark the rim of the mercury *container*. The mercury level is below LL and also below end E of the flask's neck.

increased in number and volume for about four or five days. After this, they stopped increasing and remained in exactly the same state. At the end of 12 days, once I saw that the mercury was no longer calcinating, I extinguished the fire and allowed the vessels to cool. The volume of air contained in the neck of the flask and under the empty part of the bell jar was about 50 cubic inches before the procedure, corrected for a pressure of 28 inches and a temperature of 10 degrees. After the procedure, the same air at equal pressure and temperature was found to be no more than 42 or 43 cubic inches. And so there was about a $\frac{1}{6}$ decrease in the volume of air. Furthermore, after I carefully gathered the red particles that had formed and removed as much of the liquid mercury in which they had been floating as I could, I found their weight to be 45 grains.

3.8 I had to repeat this calcination of mercury many times in closed vessels, because it is difficult, in the same experiment, to preserve both the air in which the experiment is performed and the red particles (i.e., the particles of mercury calx) that are formed. And in what follows I will often assimilate, in

[37]

3.7 *the mercury was no longer calcinating*: As a transitive verb, calcination means the operation of *roasting*. But here the verb is intransitive and denotes the process that occurs when a metal is thoroughly roasted; Lavoisier will relate the details of that process in paragraph 7.2. The substance produced by calcination of any material was commonly called a *calx*. Thus the "little red particles" that appeared on the mercury surface represent the *calx of mercury*, and he will so refer to them in paragraph 3.8. When the particles of calx stop forming, he infers that calcination has ceased. But in the case of metals, Lavoisier will eventually ban the term "calx" in favor of "oxide" (7.4).

corrected for a pressure of 28 inches and a temperature of 10 degrees: A given quantity of substance in the gaseous state does not occupy a fixed volume; rather its volume depends both on the *pressure* to which it is subjected and on its *temperature* (3.1 and comment). Comparison of gas volumes is only meaningful, therefore, if those volumes reflect identical conditions of temperature and pressure.

But gases cannot always be measured under identical conditions. Lavoisier therefore uses the gas relations to calculate what his measured volumes would be had the measurements been performed at 28 inches pressure and 10 degrees (Réaumur) temperature; he calls those calculated results the "corrected" volumes.

45 grains: The pre-revolutionary grain was equal to about .053 grams; so the quantity of calx formed in this experiment was about 2.39 grams.

3.8 *I will often assimilate ... the results of two or three experiments*: In the present instance, for example, he reported that the air volume decreased from an initial 50 to a final 42 or 43 cubic inches as 45 grains of red calx were formed. But he probably had to carry out the experiment twice: once to retrieve the calx (but letting the air escape), and again to collect the air (but perhaps destroying some of the calx).

this way, the results of two or three experiments of the same kind in the same account.

3.9 The air which remained after this procedure, and which had been reduced to $\frac{5}{6}$ of its initial volume by the calcination of mercury, was no longer fit for breathing or combustion. Animals placed in it died in a few minutes, and flames were immediately extinguished in it as if they had been immersed in water.

3.10 Furthermore, I took the 45 grains of red matter that formed during this procedure and put them in a very small glass retort to which was attached a device for receiving anything liquid or aeriform that might separate out. I lit a fire in the furnace, and I observed that as the red material was heated its color intensified. And when the retort approached incandescence, the red matter began to lose volume bit by bit, disappearing entirely in a few minutes. At the same time, about $41\frac{1}{2}$ grains of liquid mercury condensed in the little receiver, and 7 or 8 cubic inches of elastic fluid passed under the bell jar and was much more fit for combustion and animal respiration than atmospheric air.

[38]

3.11 I then collected a portion of this air by passing it into a glass tube with a one inch diameter and immersed a candle in it. The candle gave off a dazzling light. Charcoal, instead of burning quietly as it would in ordinary air, burned like phosphorus in this air, with a flame and a kind of crackling, and

---

3.10 *a device for receiving anything liquid or aeriform*: Lavoisier's descriptive phrase here lacks the completeness he usually offers. What is this "device?"

the little receiver: Unlike the vessel called by the same name in (1.16), this one is "receiver" in the original sense, a collector of the products of distillation. There is no explicit representation of it in the Treatise, but it resembles a piece of glassware drawn in Plate IV, figure 1, upon which this sketch is based. The red calx was placed in the "very small glass retort" A, joined at B to the "little receiver" C. The tube D leads to a bell jar, E, immersed in a water trough.

Initially the bell jar was filled nearly to the top with water. The heated calx gradually disappeared, while liquid mercury collected in the recipient C. At the same time, a gaseous fluid displaced the water in E, lowering its level.

3.11 *a portion of this air*: That is, a portion of the elastic fluid that accumulated in E.

Charcoal ... burned like phosphorus: Phosphorus, an ingredient in matches, burns furiously in air. Pure phosphorus can even ignite spontaneously and is then extremely difficult to extinguish.

with an intensity difficult for the eyes to bear. Priestley, Scheele and I discovered this air almost simultaneously. Priestley called it "dephlogisticated air" and Scheele called it "empyreal air." I first gave it the name "highly respirable air," but later this was changed to "vital air." We shall soon see what should be thought of these names.

3.12 On further consideration of the details of this experiment, it is clear that mercury, when it calcinates, absorbs the wholesome and respirable part of the air, or, to speak more rigorously, the base of this respirable part, and that the remaining portion of air is a kind of noxious gas, a gas which cannot

---

dephlogisticated, empyreal, vital: It is interesting to compare these names. Priestley's is a theoretical name, expressing what he took to be the absence of the supposed principle of combustion, *phlogiston*. In contrast, Scheele's and Lavoisier's adjectives express properties that can be directly observed with few, if any, extraneous suppositions. "Empyreal" (French *empiréal*) is Lavoisier's rendering of Scheele's term, "fire" (German *feuer*), and derives from the Greek ἔμπυρος, fiery; this new kind of air supports *fire* of surpassing intensity. Lavoisier's adjectives "respirable" and "vital" express the new substance's ability to maintain or support life. But the paragraph's final sentence hints at Lavoisier's dissatisfaction with *all* these characterizations, his own included.

3.12 *mercury, when it calcinates, absorbs...*: How does Lavoisier know that it *absorbed* something? Let us review the developments:

(i) the air in the jar diminished by 7–8 cu. in., leaving 42–43 cu. in. of a noxious gas residue. Presumably the 7–8 inches of lost air must have gone *somewhere*.

(ii) 45 grains of red calx formed on the mercury surface. But does that calx represent mercury to which *something has been added*, or mercury from which *something has been removed*? Even if we have reason to think the former, it is still only a guess that the *substance added to mercury*, and the *7–8 cu. in. of lost air*, are the same.

(iii) But the calx, when heated intensely, was evidently converted to 41.5 grains of liquid mercury and 7–8 cu. in. of a gas that strongly supports combustion and respiration. It therefore seems certain that the calx represents a *combination* of those two ingredients; and it cannot be mere coincidence that the 7–8 cu. in. of gas lost in the first step deprived the remaining air of respirability, while the 7–8 cu. in. of gas produced at the end exhibited exceptional respirability. Therefore "mercury, when it calcinates, absorbs the wholesome and respirable part of the air."

or, to speak more rigorously, the base of this respirable part: Recall that the gaseous state of a substance consists of (i) the substance itself, which constitutes the *base* of the gas (1.28), and (ii) sufficient caloric to maintain the particles of the base substance far enough apart from one another to erase their mutual attraction.

noxious: Harmful, poisonous, injurious, or generally unwholesome; from Latin *noxa* (injury or harm).

[39] support combustion and respiration. Thus, atmospheric air is composed of two elastic fluids of different and virtually opposite natures.

3.13 One proof of this important truth is that when these two elastic fluids that have been separated from one another (the 42 cubic inches of the noxious gas or non-respirable air, and the 8 cubic inches of respirable air) are recombined, the air that is created is in every way like atmospheric air. It supports combustion, the calcination of metals, and the respiration of animals to almost the same degree.

3.14 Although this experiment provides an extremely simple method for obtaining separately the two principal elastic fluids that constitute our atmosphere, it does not give us precise ideas of the proportion of these two fluids. The affinity that mercury has for the respirable part of the air, or rather for its base, is not strong enough to completely overcome the obstacles that impede this combination. These obstacles are the adherence of these two constitutive fluids of atmospheric air to one another and the force of affinity that unites the base of vital air to caloric. Consequently, when the calcination of [40] mercury is complete, or when, at least, it can proceed no further in a particular quantity of air, a little respirable air remains combined with noxious gas and the mercury cannot separate this last portion. I will later show that the

---

3.13 *when these two elastic fluids ... are recombined*: When he mixes the 42 cu. in. of noxious air with the 8 cu. in. of highly respirable air, the mixture has the same properties as ordinary atmospheric air.

3.14 *it does not give us precise ideas of the proportion*: It gives us an estimate, namely 8 parts respirable to 42 parts non-respirable air (a percentage of 8÷50 or 16%). But, he promises, he will presently deduce a more accurate ratio, 27:73 (equivalent to 27÷100 or 27%). Neither ratio agrees with the value presently accepted, which is 21%. But surely the non-quantitative fruits of this exercise far outweigh its quantitative deficiency! That the air we regard as our essential medium of life—upon which we depend more urgently than on food—is not life-supporting *through-and-through* but rather owes its character to a *specifically vital ingredient* is a discovery that might, in a small way, be compared to our realization that the earth is not the center of the universe.

the affinity that mercury has for the respirable part of the air ... is not strong enough: In order for mercury to capture the respirable portion of the air, it must overcome the alliance of the two parts of air with one another, and also outweigh the bond between the "base" of respirable air and the caloric that endows it with the gaseous state. If its affinity for (attraction to) the base of the respirable portion is weak, it will not fully extract the respirable part. Only at a later stage will Lavoisier be able to explain how he determined that mercury's affinity for the respirable portion of air is, in fact, too weak for that task.

ratio of respirable air to non-respirable air in atmospheric air is 27 to 73, at least in the climates we inhabit. And when I do, I will also discuss the reasons why it remains uncertain whether this ratio is precise.

3.15 Since air is decomposed when mercury calcinates, and the base of the respirable part of the air affixes to the mercury and combines with it, it follows from the principles I have previously outlined that caloric and light ought to be released. And there can be no doubt that they are. But there are two reasons that the light and caloric released will be imperceptible in the experiment I just described. First of all, the calcination lasts several days, which means that light and caloric are released over such a long interval of time that it is extremely weak at any particular moment. Second of all, because the procedure is performed with fire in a furnace, the heat generated by calcination [41] and the heat of the furnace merge. I could add that when the respirable part of the air, or rather its base, forms a new compound with mercury, it does not relinquish all the caloric combined with it, part of it remaining bound to the new compound. But this discussion, and the proofs it would require me to relate, would be out of place here.

3.16 It is, moreover, easy to make the release of heat and light perceptible, by carrying out the decomposition of air more quickly. Iron, which has much more affinity than mercury with the base of the respirable part of air, provides the means for this. By now, everyone knows the beautiful experiment of Ingenhousz on the combustion of iron. Take the end of a very fine iron thread, BC, twisted into a spiral, and attach the end B into a cork stopper, A, made for a bottle DEFG (Plate IV, Figure 17). To the other end of the iron wire, attach a small bit of tinder, C. With all of this in place, fill DEFG with air from which the non-respirable part has been removed. Light

Fig. A 17.

---

at least in the climates we inhabit: Lavoisier evidently suspects that the ratio between the respirable and the non-respirable portions of the air, whatever it may be, is not a *fixed* ratio but may depend, for example, on climate or elevation.

I will also discuss the reasons why it remains uncertain: In paragraph 3.24.

3.15 *when mercury calcinates ... caloric and light ought to be released*: When the base of respirable air combines with the mercury to form red calx, it leaves the gaseous state and must therefore expel caloric. And in paragraph 1.9 Lavoisier noted that caloric and light regularly accompany one another.

3.16 *Iron ... has much more affinity than mercury with ... the respirable part of air*: Iron shows this great affinity by its vigorous combustion in Ingenhouz's experiment, shortly to be described.

Fig. 3

Fig. 16

[42] the tinder, and then promptly plug the bottle so that the iron thread BC is inside of it as in the figure I have already cited.

3.17 As soon as the tinder is placed in the vital air, it bursts into a dazzling flame, it communicates the fire to the iron, which also burns, sending out brilliant sparks which fall to the bottom of the bottle in round droplets which turn black as they cool, and which retain a residual metallic luster. The iron burned in this way is even more brittle and fragile than glass. It is easily reduced to a powder, and magnets still attract it, though less so than before combustion.

3.18 Ingenhousz did not examine what happened to the iron or to the air in this procedure, so I found myself obliged to repeat the experiment under different conditions and with equipment better suited to my purposes.

3.19 I filled a bell jar, A (Plate IV, Figure 3), with about 6 pints of pure air, that is to say, with the "highly respirable" part of the air. I carried this

---

3.17 *the iron ... burns, sending out brilliant sparks*: Such vehement combustion indicates a strong affinity of iron for the base of vital air. If mercury cannot similarly be set aflame in purely vital air, one would conclude that its affinity for the respirable part of air is weaker (3.16).

3.19 *I filled a bell jar ... with about 6 pints of pure air*: An unfilled jar already contains ordinary air, so how can he fill it with pure air? He first fills the jar with water by immersing it in a tub of water. Then, holding the jar mouth down as shown in the drawing, he uses a suitably bent tube to bubble pure air into the jar; the pure air replaces the water in the jar.

pure ... "highly respirable": Lavoisier is still casting about for a suitable name. He will propose one in the following chapter, after having discussed the principles that will govern his naming of newly-discovered substances.

bell jar over to a bath of mercury contained in a basin, BC, with the aid of a very shallow bowl, after which I carefully dried the surface of the mer- [43] cury with scraps of blotting paper, both inside and outside the bell jar. Furthermore, for the purposes of this experiment, I used a little porcelain saucer, both flat and flared at the lip, in which I had placed iron shavings twisted into spirals and arranged in a manner which seemed to me most likely to allow the combustion to spread to all of the chips. I attached a bit of tinder to the end of one iron chip, and to this I added a fragment of phosphorus which hardly weighed $\frac{1}{16}$ of a grain. I placed the saucer under the bell jar by lifting the end a little. I realize that when the experiment is performed this way, a little bit of common air mixes with the air under the bell jar, but when one works quickly, this mixing in no way compromises the experiment's success.

3.20 Once I had placed the saucer D under the bell jar, I sucked out part of the air from the bell jar in order to raise the mercury inside it to the

---

I carried this bell jar over to a bath of mercury ... with the aid of a very shallow bowl: While the mouth of the jar (now filled with pure air) is still under water, he slides the bowl beneath it, as in the drawing. When he lifts both jar and bowl out of the tub, enough water remains in the bowl to seal the opening and thus retain the pure air in the jar. He carries both jar and bowl to a bath of mercury, immerses them in the mercury bath, and removes the bowl. His Figure 3 depicts the result.

I carefully dried the surface of the mercury: Since water is less dense than mercury, the water brought over in the shallow bowl inevitably floats upon the mercury surface, both inside and outside the jar. He uses blotting paper, which soaks up water but not mercury, to dry it. How might he manipulate the blotting paper so as to dry the mercury *within* the jar?

Why could he not have filled the bell jar over mercury at the outset, and so avoid introducing water into the mercury apparatus at all? The tub into the bell jar was immersed would then have had to contain mercury, which is nearly fourteen times heavier than the same quantity of water. Lavoisier indicates in Part III of the Treatise that such weight was more than any wooden tub could support.

I added a fragment of phosphorus: Phosphorus is strongly disposed to ignite even in common air, much more so in highly respirable air. Along with the "bit of tinder," it serves as a fire-starter to help kindle the iron shavings.

3.20 *Once I had placed the saucer under the bell jar:* Saucer D, containing the iron shavings, *floats* on the mercury surface.

height EF. A siphon, GHI, should be used for this purpose. It should be plugged with a wad of paper to keep it from filling with mercury, and then inserted into the bell jar from below. There is an art to sucking air in such a way as to raise the mercury under a bell jar. If you just use your lungs, you can only raise it a little, perhaps an inch, an inch and a half at most. But if you also use the muscles of your mouth, it is possible to raise the mercury about 6 or 7 inches without tiring yourself, or at least without too much inconvenience.

[44]

3.21 After all of this was prepared, a bent piece of iron, MN, (Figure 16) designed for experiments like these, is heated until it glows red. It is then inserted into the bell jar from below; and before it has had time to cool, it is brought close to the bit of phosphorus contained in the porcelain saucer, D. As soon as the phosphorus catches fire, it communicates its flame to the

---

3.20, cont. *It should be plugged with a wad of paper to keep it from filling with mercury*: The plugged end I of the siphon is passed under the jar, through the mercury, and into the highly respirable air above the mercury surface. Remarks that Lavoisier will make in paragraph 3.24 indicate that the plug is not airtight; he closes end G of the tube with a finger while the plugged end is immersed.

use the muscles of your mouth: In using the mouth rather than the lungs to create suction, Lavoisier would have reduced only slightly the danger of inhaling toxic mercury vapor. Such considerations were surely no part of his thinking; but the technique is a risky one. Today, anyone working with mercury would use a mechanical aspirator to perform that task.

3.21 *before it has had time to cool*: One end of the iron wire is grasped with a suitable tool; the other end is plunged into the mercury bath, then maneuvered under the jar edge and raised so as to contact the phosphorus. If this is done quickly, the iron will still be hot enough to ignite the phosphorus.

tinder, which in turn communicates it to the iron. When the iron chips have been arranged well, a brilliant white flame, like Chinese fireworks, spreads out over the iron and burns every last bit. The great heat produced during this combustion liquefies the iron, which falls in round drops of different sizes. These drops stay in the saucer for the most part, but some jump out and float on the surface of the mercury.

3.22 During the first moment of combustion, there is a slight increase in the volume of the air because of the expansion caused by the heat. But rapid contraction soon follows this expansion. The mercury rises again in the bell jar, and when there is enough iron and the air used is very pure, almost all of the air is absorbed. [45]

3.23 So I should warn readers here that unless one is making experiments for the purposes of research, it is better to burn only modest quantities of iron. For when the experiment is pushed too far and almost all the air is absorbed, the saucer D which floats on the mercury comes too close to the top of the bell jar, and the great heat, together with the sudden cooling caused by contact with the mercury, shatters the glass. And as soon as there is a crack in the bell jar the weight of the column of mercury causes it to fall very quickly, and this leads to such a flood, that most of the mercury gushes out of the basin. In order to avoid these inconveniences, and to be sure that the experiment succeeds, no more than 1.5 gros of iron should be burned under a bell jar with a volume of 8 pints. And this bell jar should be strong enough to hold up against the weight of the mercury it will contain.

3.24 In this experiment, it is impossible to determine the weight the iron acquires and the changes that occur in the air at the same time. If the aim is to determine the weight of the iron and its relationship to the absorption of air, be sure to mark the height of the mercury on the bell jar very exactly with the tip of a diamond both before and after the experiment. Then pass the siphon, GHI, which has been plugged with paper to prevent it from filling with mercury, under the bell jar. Place a thumb on one end, G, and then allow air in under the thumb a little at a time. When the mercury has descended [46]

3.22 *when ... the air used is very pure*: That is, when it contains the respirable part of air almost exclusively.

almost all of the air is absorbed: The same reasoning that implied "absorption" of vital air by mercury (3.12) will apply to the iron shavings used here; and, evidently, a very small quantity of iron is capable of "absorbing" nearly 6 pints (3.19) of vital air.

3.23 *no more than 1.5 gros of iron should be burned*: 1.5 gros is equivalent to 5.7 grams. For this and other equivalents, see the table of pre-revolutionary weights and measures in the Appendix.

to its level, slowly lift the bell jar, remove the droplets of iron from the saucer, and carefully gather any that have splashed out and are floating on the surface of the mercury. Then weigh all the iron. This iron is in a state that earlier chemists called "martial ethiops." It has a kind of metallic luster, is very brittle, crumbles easily, and can be reduced to

a powder with mortar and pestle. When the procedure works well, 135 or 136 grains of martial ethiops can be obtained from 100 grains of iron. So the increase in weight is at least 35 parts per hundred.

[47]

3.25 If this experiment has been carried out with the careful attention it deserves, the weight of the air diminishes by an amount exactly equal to the increase in the weight of the iron. So if 100 grains of iron are burned and the iron gains 35 grains, the volume of air will diminish, at a rate of half a grain per cubic inch, by almost exactly 70 cubic inches. We will see in later chapters of this memoir that the weight of vital air is in fact almost exactly half a grain per cubic inch.

3.26 I will remind readers here for the last time, that in all experiments like

---

3.25 *the weight of the air diminishes by an amount exactly equal to the increase in the weight of iron*: Therefore the total quantity of matter (evidenced by its weight) involved in the combustion is *constant*. This is a general principle for Lavoisier, which, remarkably, he does not state explicitly until Chapter Thirteen. There he writes: "For nothing can be created by either natural or artificial processes. The amount of matter at the outset of any process is equal to the amount of matter that remains afterwards..." Lavoisier was by no means the first to propound the principle of *conservation of mass*, as it is called today. But he was the first to treat it as not just a philosophical truism but a practical tool of analysis. Scrupulous determination of weights, using a superbly-crafted balance, made this possible. The photo shows the analytical balance made for Lavoisier by Fortin in 1788, in the collection of the Musée des Arts et Métiers, Paris. Used by permission.

half a grain per cubic inch: This is Lavoisier's figure for the density of highly respirable air at the standard (for him) pressure and temperature of 28 inches of mercury and 10° Réaumur. Lavoisier reserves the description of his methods for determining weights and volumes of gases for Part III of the Treatise.

this one, it is necessary to recalculate the volume of air, both at the beginning and at the end of the experiment, in order to determine what it would have been at a temperature of 10 degrees and a pressure of 28 inches. I will explain in some detail how to make the necessary corrections at the end of this work.

3.27 If the aim of the experiment is rather to investigate the quality of the air that remains in the bell jar, a somewhat different approach is required. After the combustion is complete and the vessels have cooled down, begin by removing the saucer that contains the iron by passing your hands under the bell jar and through the mercury. Then place, under the same bell jar, some potash or caustic alkali dissolved in water, or some sulphuret of potash or some other such substance as appropriate, in order to examine the effect that these substances exert on the air. I will return again to these other [48] methods for analyzing air once I have explained the nature of the different substances which I mention only incidentally here. Finally, introduce enough water into the bell jar to displace all the mercury, and pass a vessel such as a very flat saucer under the bell jar so that it can be carried over to a common pneumato-chemical apparatus for water, with which it is possible to work on a larger scale and with greater facility.

3.28 When very soft, very pure iron is used, and when the combustion takes place in respirable air without any mixture of non-respirable air, the air that remains after combustion will be found to be as pure as it was before combustion. But it is rare for the iron not to contain a small quantity of charcoal matter. Steel always contains some. It is also extremely difficult to obtain perfectly pure respirable air. This air is almost always mixed with a little bit of non-respirable air, but this kind of noxious gas in no way compromises the results of the experiment, and at the end of the experiment the same amount of it that was there in the beginning remains.

3.29 Earlier, I indicated that there are two ways to determine the nature [49] of the constituents of atmospheric air, decomposition and composition. The calcination of mercury furnishes us with an example of both, since, after

---

3.26 *how to make the necessary corrections*: Lavoisier's explanation appears in Part III of the Treatise. His procedure is essentially equivalent to the modern application of Boyle's (or Marriotte's) and Charles' laws (3.1, comment).

3.27 *pneumato-chemical apparatus*: The name properly refers to an elaborate device Lavoisier developed for separating mixed gases. But here he seems to mean only the *pneumatic trough*, which enables collection of gases in a bell jar. Lavoisier's Plate IV, figure 3 (on the previous page) depicts a small example, suitable for collecting gases over mercury. Lavoisier also employed a larger version for use with water; such was the tub referred to in (3.19, comment). The actual "pneumato-chemical apparatus" indeed included a pneumatic trough, but also much more (13.7, comment).

having removed the base from the respirable part of the air with mercury, we replaced it so as to form air in every way like atmospheric air again. But it is equally possible to affect the composition of atmospheric air by gathering the required materials from different kingdoms of nature. We will see below that dissolving animal matter in nitrous acid releases a great quantity of the kind of air that extinguishes flames, air that is harmful to animals, and which is in every way like the non-respirable part of atmospheric air. If 27 parts of highly respirable air, which has been extracted from red calx (i.e., calcinated mercury), are added to 73 parts of this non-respirable air, an elastic fluid exactly like the atmosphere and with all the same properties is formed.

3.30 There are many other ways to separate the respirable from the non-respirable part of the air, but I could not explain them without appealing to concepts which, in the order of knowledge, belong to later chapters. The experiments I have already reported suffice for an elementary treatise, and in these sorts of matters, it is more important to choose proofs carefully than to present a great number of them.

[50]

3.31 I will conclude this chapter by mentioning a property of the atmosphere which, in general, all elastic fluids or gases that we know of share: that of dissolving water. A cubic foot of atmospheric air can dissolve about 12 grains of water, according to the experiments of Saussure. Other elastic fluids, such as carbonic acid, can dissolve more, but the experiments to determine exactly how much have not yet been made. The water that elastic, aeriform fluids contain gives rise to certain phenomena in some experiments which merit more attention and which have often thrust chemists into great errors.

3.29 *nitrous acid*: Lavoisier will discuss, and name, this acid in Chapter Six.

exactly like the atmosphere: Since (i) the combination has the same properties as common air, and (ii) one of the ingredients is known at the outset to be the respirable part of common air, then (iii) the second ingredient would appear to be identical to the non-respirable part of the air—even though it was obtained from a source that has no obvious connection to the atmosphere.

3.31 *dissolving water*: The most common cases of "dissolving" are those that feature a solid or liquid *solute* (sugar or milk) and a liquid *solvent* (coffee or tea). But Lavoisier does not hesitate to extend ordinary usage to include the case of a liquid solute (water) and a gaseous solvent (air). And he has already compared caloric— the paradigmatic elastic fluid—to a solvent (1.28).

carbonic acid: This acid will be discussed in Chapter Five and named in Chapter Six.

great errors: One such error can arise when gases are collected over water in the pneumatic trough. The desired gas will inevitably harbor some water vapor, which will lead to error if not taken into account.

Chapter Four

A Nomenclature for the Different Constituent Parts of Atmospheric Air

4.1 So far, I have been forced to use periphrases to designate the nature of the different substances which compose our atmosphere. I have provisionally adopted expressions such as "the respirable part of the air" and "the non-respirable part of the air." But I am about to go into much greater detail, and this calls for a more rapid pace. I have already tried to convey simple ideas of the different substances which go into the composition of atmospheric air. From now on, I will also use simple words to refer to them.

4.2 The temperature of the planet we inhabit is found to be very close to the degree at which water passes back and forth between a liquid and a solid state. This phenomenon frequently occurs before our very eyes, and so it is not surprising that in all languages, at least those spoken in climates where they have winter, there is a name for water that has become solid through lack of caloric.

4.3 But the same cannot be said for water converted to a vaporous state through the absorption of a greater quantity of caloric. Those who have not [52] made a special study of these matters still do not know that, at a temperature slightly higher than that at which it boils, water transforms into an elastic, aeriform fluid which, like all gases, can be captured and contained in vessels, and which keeps its gaseous form so long as it is exposed to temperatures greater than 80 degrees under a pressure equal to that of a 28 inch column of mercury. Since this phenomenon has escaped the notice of most people, no language has a word for water in this state, and it is the same for all fluids and, in general, for all substances which never evaporate at what are for us ordinary temperatures and pressures.

4.2 *there is a name for water that has become solid through lack of caloric*: In English that name is, of course, *ice*.

4.3 *no language has a word for water in this state*: One might think that both the French *"vapeur"* and English "steam" are such words. But the "state" Lavoisier is speaking of is a pure, invisible gas; whereas, in popular use, both these terms refer to substances like the exhalations of kettles, which form visible clouds because they contain minute droplets of liquid water. Since the eighteenth century, however, both terms have acquired Lavoisier's meaning in scientific and technical contexts.

4.4 For the same reason, there is no word for most aeriform fluids in the liquid or solid state. For no one knew that these fluids were the result of the combination of a base with caloric, and since they had never been seen in either a liquid or a solid state, no one, not even natural philosophers, knew that they existed in these forms.

4.5 We decided that it was not open to us to change conventional names which had been thoroughly established in society through long use. For this [53] reason we have retained the colloquial senses of the words "water" and "ice." We have even used the word "air" to signify the collection of elastic fluids that compose our atmosphere. But we did not consider ourselves similarly obliged to adhere to the very modern names recently proposed by natural philosophers. We thought instead that we were right to reject these names and replace them with others less likely to lead to errors. Even when we decided to adopt recently proposed names, we had no qualms about modifying them or attaching them to more distinct and restricted ideas.

4.6 We have drawn these new words principally from the Greek, and we have chosen them so that their etymologies call to mind the ideas of the things we intend to signify. Above all else, we were committed to using short words and, whenever possible, short words from which adjectives and verbs could be formed.

4.7 In accordance with these principles, we have followed the example of Macquer and have employed the name "gas" used by van Helmont; and we have subsumed the large class of elastic, aeriform fluids under this name, with [54] the exception of atmospheric air. Thus, the word "gas" as we use it is a generic name which designates the last degree of saturation of some substance by caloric; it is a word that expresses a mode of corporeal being. It was then necessary to find a specific name for each kind of gas, and we did this by deriving a second word from the name for the base of each gas. Thus, we will call water combined with caloric in the state of an elastic, aeriform fluid "aqueous gas." We will call ether combined with caloric "ethereal gas," spirit of wine combined with caloric, "alcoholic gas," and similarly for "muriatic acid gas," "ammoniac gas," and all the others. I will further expand this list of names for different gases when I provide the names for their different bases.

4.4 *combination of a base with caloric*: Recall Lavoisier's usage of "base," discussed in the comment to paragraph 1.28.

4.7 *a mode of corporal being*: That is, a *state* or *condition* of body, rather than a fixed property.

4.8 We have seen that atmospheric air is principally composed of two aeriform fluids or gases. One is respirable and capable of sustaining the life of animals. Metals calcinate in it, and combustible bodies burn. The other has absolutely opposite properties: animals cannot breathe in it, it cannot sustain combustion, and so on. We named the base of the respirable part of the air "oxygen," which we derived from two Greek words: ὀξύς, ("acid") and γείνομαι ("I generate") because one of the most common properties of this base is to form acids when it combines with most substances. So when this base unites with caloric we will call the result "oxygen gas." Its weight in this state is almost exactly half a standard grain per cubic inch or an ounce and a half per cubic foot at a temperature of 10 degrees and a barometric pressure of 28 inches.

[55]

4.9 Since the chemical properties of the non-respirable part of atmospheric air are still not well known, we were content to derive the name of its base from this gas's property of killing animals that breathe it. We have for this reason called it "azote," using the Greek alpha privative and the word ζωή, "life." And so the non-respirable part of the air will be called "azotic gas." Its weight is 1 ounce, 2 gros, 48 grains per cubic foot or 0.4444 grains per cubic inch.

4.10 We were not unaware that this name seemed strange, but this is the fate of all new names, and it is only through use that we become accustomed to them. We spent quite some time exploring alternatives, but it proved

---

4.8 ὀξύς (acid): The Greek adjective means *sharp, acute,* or *pungent.* Lavoisier will demonstrate the role of oxygen as "acid-maker" in Chapters Five and Six.

4.9 *Since the chemical properties ... are still not well known...*: Ideally, the name of a simple substance should reflect its chemical properties; when these are unknown, other properties must serve as the basis for naming. But what distinguishes "chemical" properties from other kinds of properties?

...we were content to derive the name [azote] from this gas's property of killing animals that breathe it: Evidently Lavoisier does not regard azote's *inability to support respiration* as a chemical property, in contrast to oxygen's *ability to form acids*, which he does view as "chemical." Now the ability of one substance to form another (as oxygen does) is a relation between a substance and a substance; but the inability of a azote to sustain life is a relation between a substance and an *entire complex organism.* Perhaps, then, a "chemical" property is one that acts at the substance-to-substance level.

the Greek alpha-privative: Numerous Greek words express the negation of something by prefixing the letter *alpha* to its name. English words like *atypical* (meaning "not typical") and other words derived from Greek carry over this formation.

impossible to find a better name than this one. We were at first tempted to call it "alkaligenic gas" in recognition of Berthollet's experiments which proved that this gas is a constituent of volatile alkali (ammoniac), as will be seen in what follows. But on the other hand, we have as yet no proof

whatever that it is one of the constituent principles of other alkalis, and, furthermore, it had been shown to be a constituent of nitric acid, and so we had equal grounds for calling it "the nitrigenic principle." In the end, we had to reject the name, though it conveyed an idea in a manner consistent with our system; and we thought we avoided the risk of making a mistake by adopting the names "azote" and "azotic gas," since these only express a fact, or rather a property that the gas possesses, namely the property of killing the animals that breathe it.

4.11 If I said anything more about the nomenclature of the different kinds of gases, I would need to appeal to notions reserved for later chapters. So I will not provide names for all the gases here. For now, it suffices that I have explained the method used to name them all. The chief merit of the system of nomenclature that we have adopted is that the names of all of the compounds of a particular simple substance follow necessarily from that simple substance's name.

4.10 *alkaligenic*: Such a name would express a genuinely chemical property, the *ability to generate alkalis*; it would, therefore, be preferable to "azote." But what *are* alkalis? We will learn more about this class of substances in Chapter Sixteen.

volatile alkali (*ammoniac*): This substance, which Lavoisier mentioned earlier in paragraph 1.28, is an example of an alkali that contains azote. But since it has so far proved to be the *only* example, we cannot regard the *ability to generate alkalis* as characteristic of azote.

we had to reject the name, though it conveyed an idea in a manner consistent with our system: The name "alkaligenic," expressing *ability to generate alkalis*, would have been "consistent with our system," since it would represent a chemical property—a capacity of the substance in question to act with respect to other substances (0.32). But since that power could be confirmed for azote only in a single instance and not generally, the name had to be discarded. A similar consideration disqualifies "nitrigenic."

"azote" and "azotic gas" ... only express a fact, or rather a property that the gas possesses: The name doesn't express a *chemical* property, which would be preferable. But at least it expresses a *fact*; that is, it asserts only so much as we may observe.

Chapter Five

On the Decomposition of Oxygen Gas by Sulphur, Phosphorus, and Charcoal, and on the Formation of Acids in General

5.1 One principle of the art of experimentation, a principle of which one should never lose sight, is that experiments should be simplified as much as possible; that every circumstance that could complicate the results should be eliminated. Thus, in the experiments treated in this chapter, atmospheric air will not be used, since it is by no means a simple substance. It is certainly true that azotic gas, a part of the mixture that constitutes air, appears to be entirely passive in calcinations and combustions. But since it slows down these processes, and it is not impossible that it even changes their outcomes in certain cases, it seemed to me necessary to eliminate this source of uncertainty.

5.2 Therefore I will only present the results of combustions that took place in vital air (i.e., in pure oxygen gas) in the experiments described below, and I will only note those differences which occur when oxygen gas is mixed with azotic gas in different proportions.

Fig. 3

5.3 I took a crystal bell jar, A, with a volume of between five and six pints (Plate IV, Figure 3). I filled it with oxygen gas over water, after which I transported it to a bath of mercury by means of a glass saucer that I placed beneath it. I then dried the surface of the mercury and introduced, under the bell

---

5.3 *I filled it with oxygen gas over water.... I then dried the surface of the mercury*: Lavoisier again uses the filling procedure discussed earlier in the comment to paragraph 3.19. As before, that technique introduces some water into the apparatus and therefore requires him to dry the mercury surface.

jar A, $61\frac{1}{4}$ grains of Kunkel's phosphorus on two porcelain saucers similar to the one depicted in D, Figure 3. I covered one of the two portions of phosphorus with a small plate of glass, so that I could light each of the two portions separately, and so that the flame would not pass from one to the other. Once all of this had been arranged, I raised the level of mercury in the jar to height EF by sucking on a glass siphon GHI that had been inserted beneath the jar. In order to avoid filling the siphon as it passed through the mercury, I plugged one end with a small wad of paper. Then, with a piece of curved wire (as depicted in Plate IV, Figure 16) that had been heated in a fire until it glowed red, I lit the two saucers of phosphorus successively, beginning with the one which was not covered by the square of glass.

[59]   5.4 The combustion occurred very rapidly. There was a brilliant flame, and a considerable amount of heat and light was released. In the first instant, there had been a considerable expansion of oxygen gas due to the heat. But soon afterwards the mercury rose back above its level, and a considerable amount of oxygen was absorbed. At the same time, the whole interior of the jar was covered with light, white flakes, which were nothing other than solid phosphoric acid.

5.3 *Kunkel's phosphorus*: Johann Kunkel, though not its discoverer, produced phosphorus commercially by a proprietary process. The quantity $61\frac{1}{4}$ grains is very nearly equal to 3.25 grams.

with a piece of curved wire ... heated in a fire until it glowed red, I lit the two saucers of phosphorus: Lavoisier used this technique earlier in paragraph 3.21.

5.4 *a considerable amount of oxygen was absorbed*: Since the mercury "rose back above its level," it is clear that the total amount of oxygen in the jar decreased. By analogy with his earlier combustion of mercury (3.5) and iron (3.19), it is reasonable to suppose that the vanished oxygen had been absorbed by the phosphorus. But he will, as always, confirm this supposition by weighing the materials involved.

light, white flakes, which were ... solid phosphoric acid: Lavoisier is anticipating here, as

5.5 After all corrections had been made, there were 162 cubic inches of oxygen gas at the beginning of the experiment. At the end, there were only $23\frac{1}{4}$ cubic inches. Thus $138\frac{3}{4}$ cubic inches or 69.375 grains of oxygen gas had been absorbed.

5.6 Not all of the phosphorus burned. Several pieces remained in the saucers. When these were washed so as to separate out the acid, and then dried, they were found to weigh approximately $16\frac{1}{4}$ grains. Thus 45 grains of phosphorus, more or less, were burned. I say "more or less" because it is possible that there was an error of one or two grains with respect to the weight of the phosphorus remaining after combustion.

5.7 Thus in this procedure, 45 grains of phosphorus combined with 69.375 grains of oxygen. And since nothing with weight passed through the glass, we are entitled to conclude that the weight of whatever substance was produced by this combination, and which had collected in white flakes, should be the same as the sum of the weights of the oxygen and the phosphorus, that is, [60]

---

indeed he anticipated earlier (4.8) when he stated that oxygen possesses the ability to generate acids. Not until paragraph 5.17 will he actually show that these white flakes have an acidic character.

5.5 *After all corrections had been made*: That is, after converting the gas volumes measured under actual conditions to the volumes those same quantities of gas would have under Lavoisier's standard conditions of 28 inches pressure and temperature of 10° Réaumur.

$138\frac{3}{4}$ cubic inches or 69.375 grains of oxygen gas had been absorbed: $138\frac{3}{4}$ is the difference between the corrected initial volume of oxygen and the corrected final volume. And in paragraph 4.8, Lavoisier reported the weight of a cubic inch of oxygen (at his standard pressure and temperature) to be almost exactly one-half grain. Multiply 138.75 by .5 to obtain Lavoisier's figure.

5.6 *these were washed to separate out the acid*: Unburned pieces of phosphorus were coated with some of the white flaky material. As Lavoisier will remark in paragraph 5.17, the white substance is highly soluble in water (indeed, such solubility is one of the typical properties of acids). Phosphorus, on the other hand, does not react with water; so washing is practical. After washing and drying, the unburned phosphorus could be weighed.

45 grains of phosphorus ... were burned: 45 is the difference between the initial weight of phosphorus, $61\frac{1}{4}$ grains, and the remaining weight, $16\frac{1}{4}$ grains.

5.7 *nothing with weight passed through the glass*: It is of course certain that *caloric* passed through the glass (5.4); but caloric has no detectible weight.

the weight ... produced ... should be ... the sum of the weights of the oxygen and the phosphorus: Note this implicit use of the principle of constancy of weight, discussed in (3.25, comment).

114.375 grains. We will soon see that the white flakes are nothing other than a solid acid. Expressing these quantities in parts per hundred, we find that 154 livres of oxygen are required to saturate 100 livres of phosphorus, and this produces 254 livres of white flakes, i.e., solid phosphoric acid.

5.8 This experiment proves in a very clear way that oxygen has more affinity for phosphorus than for caloric at a certain degree of temperature. Consequently, phosphorus decomposes oxygen gas, attaches to its base, and then the caloric which is liberated escapes and disperses as it spreads into surrounding bodies.

5.9 But no matter how conclusive this experiment may have been, it was not yet sufficiently rigorous. In fact, it is not possible to verify the weight of the white flakes (i.e., the solid acid) produced in the apparatus that I used and just described. We can only infer this weight by means of a calculation, by assuming that it is equal to the sum of the weights of the oxygen and phosphorus. But however evident this inference may be, in chemistry and physics one must never infer something that can be ascertained through

[61]

5.7, cont. *Expressing these quantities in parts per hundred*: 69.375 is to 45 as 154 is to 100.

5.8 *oxygen has more affinity for phosphorus than for caloric...*: Since the oxygen was initially in gaseous form, it must have been united to caloric. But when it combined with the phosphorus it became a solid, thereby giving up considerable caloric. Oxygen must therefore have been attracted to phosphorus more than to caloric.

...at a certain degree of temperature: The phosphorus had to be ignited—that is, brought to a sufficiently high temperature—before oxygen would combine vigorously with it. But phosphorus's tendency to *glow spontaneously* in air suggests that oxygen may have considerable affinity for it even at ordinary temperatures.

phosphorus decomposes oxygen gas: Modern chemistry teaches that oxygen is an elementary substance. And even Lavoisier has no evidence to suggest that it is other than "simple"; how then can it be "decomposed"? Recall that, for Lavoisier, every gas is a compound of that gas's *base* and *caloric*. Thus any gas that becomes either liquid or solid undergoes decomposition.

5.9 *It is not possible to verify the weight of the white flakes*: Notice that Lavoisier made no effort to collect the white flaky material for weighing. Even had he done so, the operation of washing the unburned phosphorus to remove the white coating (5.6) would have caused the loss of an undetermined amount of the white substance.

one must never infer something that can be ascertained through direct experiment: Here, Lavoisier is applying this maxim to his doctrine that *no substance is either created or destroyed* (3.25, comment)—a doctrine that is crucial to his judgement that oxygen was *absorbed* by the phosphorus (5.4). Even though he may have had direct experiment. I therefore believed it was necessary to perform the experiment again on a somewhat larger scale and with different equipment.

5.10 I took a large spherical glass flask, A, with an opening, EF, three inches in diameter (Plate IV, Figure 4). I covered this opening with a crystal plate deburred with emery paper and into which two holes had been drilled so as to allow the tubes *xxx* and *yyy* to pass.

Fig. 4.

5.11 Before closing up the flask with its plate, I set up a stand, BC, inside of it, on top of which was a porcelain saucer, D, containing 150 grains of phosphorus. When everything had been arranged in this way, I placed the crystal plate over the opening of the flask, and I sealed the opening with a lute made from fat and covered the seal with linen bands soaked in quicklime and egg whites. When the lute was thoroughly dry, I suspended the whole apparatus from the arm of a balance, and I determined its weight within nearly a grain or a grain and a half. Next I fitted tube *xxx* to a small air pump and evacuated the flask. Then I let oxygen gas into the flask by opening a stopcock fitted to tube *yyy*. I will note here that experiments like these can be performed easily enough and,

[62]

---

good reason to suppose such absorption by analogy with earlier experiments on combustion, he will repeat the experiment using apparatus that enables him to weigh all the materials involved.

5.10 *a crystal plate deburred with emery paper*: Lavoisier would have cut or sawn the circular crystal (glass) plate EF, whose edges would then require smoothing and sanding. Emery paper is coated with a hard, impure, abrasive mineral, and was used the way "sandpaper" is used today.

5.11 *lute*: A clay or cement for sealing joints; from Latin *lutum*, mud.

quicklime: It becomes a powerful alkali when mixed with water, but here Lavoisier is using it for its sticky, adhesive qualities. Compare Wordsworth: "...a snare for inattentive fancy, like the lime that foolish birds are caught with" (*Sonnet composed upon Westminster Bridge, Sept. 3, 1802*).

I determined its weight: Lavoisier did not do this in the first experiment. Weighing the flask *and its contents* allows him to monitor the total weight of materials involved.

evacuated the flask: He removed from the flask as much common air as possible, in order to replace it with pure oxygen gas.

more importantly, with great precision using the hydro-pneumatic machine Meusnier and I described,* and which is explained in the last part of this work. By means of this instrument, to which Meusnier has since made important additions and improvements, one can ascertain precisely both the quantity of oxygen gas introduced into the flask, and the amount consumed over the course of the procedure.

    5.12 Once I had set up the apparatus in the way described above, I set fire to the phosphorus with a burning glass. The combustion was extremely rapid, accompanied by a great flame and much heat. During combustion, a great quantity of white flakes formed, which attached themselves to the inner walls of the vessel, soon enshrouding it completely. The vapors emitted were so profuse that even though the combustion ought to have been sustained by the new supplies of oxygen gas continually passing into the flask, the phosphorus was soon extinguished. Once I had allowed the whole apparatus to cool off completely, I began by determining the quantity of oxygen gas that

[63]

Mémoires de l'Académie, 1782, p. 466.

5.11, cont. *hydro-pneumatic machine*: Also called *gasometer*, this device delivered a measured volume of gas at constant pressure. Since Lavoisier has already determined the weight of one cubic inch of oxygen (3.25), he can calculate the weight of any measured volume. The gasometer is convenient for the present experiment; it will be indispensable for an experiment to be described in Chapter Eight.

5.12 *burning glass*: Lavoisier used a large magnifying glass to focus the sun's rays on the phosphorus, thereby heating it to ignition.

    the new supplies of oxygen gas continually passing into the flask: The hydro-pneumatic machine automatically supplied oxygen to the flask in whatever quantity was needed to keep the pressure constant. In so doing, it both replaced and measured the amount of oxygen consumed in combustion.

    to cool off completely: He allowed the apparatus to regain, as nearly as possible, the temperature it had at the start of the experiment. The hydro-pneumatic machine remained connected, thereby keeping the pressure constant during this period.

    I began...: That is, he began the process of *assessing the results* by carrying out two operations: (i) measuring the amount of oxygen consumed, and (ii) weighing the flask and its contents, including the newly-formed white flakes.

    ...by determining the quantity of oxygen gas that had been used...: This quantity was measured by the hydro-pneumatic machine (5.11). The machine kept the oxygen pressure constant, and after combustion the oxygen returned to its initial temperature. Then since the conditions before and after combustion were identical, the quantity of oxygen supplied must exactly equal the quantity of oxygen absorbed.

had been used, and by weighing the flask before opening it up. I then washed, dried, and weighed the small quantity of phosphorus which remained in the saucer and which was yellow in color like ochre, so as to deduce the total quantity of phosphorus used in the experiment. It is clear that with the help of these different precautions, it was easy for me to ascertain (1) the weight of phosphorus burned; (2) the weight of the white flakes produced by combustion; and (3) the weight of oxygen gas which had combined with the phosphorus.

5.13 This experiment gave me more or less the same results as the preceding one. Just as before, when the phosphorus burned, it absorbed a little more than one and a half times its weight in oxygen; and in addition I became more certain that the weight of the new substance produced was equal to the sum of the weights of the burned phosphorus and the oxygen it had absorbed, which was, in any case, easy to foresee *a priori*.

...and by weighing the flask before opening it up: He again weighs the flask and its contents, as he did prior to the combustion. Is it obvious what purpose is served by this second weighing?

I then washed, dried and weighed the ... phosphorus which remained: The difference between the initial weight of phosphorus and the weight that remained after combustion is the weight of phosphorus that was burned.

ochre: An earth rich in clay, with color varying from yellow to brown; it is widely used as a pigment.

precautions: On its face, a strange word to use here. A *precaution* is a special measure taken in advance in order to guard against a known risk. But washing and drying the residual phosphorus is merely routine procedure prior to weighing it.

Not so, however, for weighing the flask! Since the total quantity of matter involved in the combustion is constant (3.25), the difference between the final and initial weights of the *flask with its contents* must be equal to the *weight of oxygen supplied*. But Lavoisier made it clear in paragraph 5.11 that he relied on the *hydro-pneumatic machine* to measure the oxygen. Weighing the flask, therefore, appears indeed to be a genuine "precaution," providing an alternative determination of the oxygen consumed; if the two measures fail to agree, that will alert Lavoisier to an error in either his reasoning or his procedure. Compare Lavoisier's phrase "double-check" in (8.28)

it was easy for me to ascertain...: He has both (1) the weight of phosphorus burned, and (3) the weight of oxygen that combined with it, by direct measurement. Then since the total quantity of matter involved in the combustion is constant (3.25), the weight of the white flakes (2) is simply the sum of (1) and (3).

5.13 *a priori*: A logical term, here meaning "in advance."

5.14 If the oxygen gas used in this experiment is pure, the residue that remains after combustion is equally pure. This proves that nothing is emitted from the phosphorus that can compromise the purity of the air, and that phosphorus does nothing else but remove the base, which is oxygen, from the caloric to which it is united.

[64]

5.15 I mentioned above that if a combustible body were burned in a hollow sphere of ice or in any apparatus constructed according to the same principle, the quantity of ice melted during combustion would be an exact measure of the quantity of caloric released.* When we tested the combustion of phosphorus in this way, we discovered that burning 1 livre of phosphorus melted a little more than 100 livres of ice.

5.16 The combustion of phosphorus succeeds equally well in atmospheric air, with only these two differences: (1) the combustion is much less rapid, since it is slowed down by the large proportion of azotic gas which is mixed with oxygen gas; (2) only a fifth of the air at most is consumed, since only the oxygen gas is consumed, and so the proportion of azotic gas becomes so great towards the end of the procedure that combustion can no longer take place.

[65]

5.17 As I already mentioned, phosphorus, whether it burns in ordinary air or in oxygen gas, is transformed into a very light, flaky material, acquiring completely new properties: though it is initially insoluble in water, it not only becomes soluble, but it draws the moisture from the air with astonishing

---

* For more details see the memoir that Laplace and I jointly presented (*Mémoires de l'Académie*, 1780, p. 355).

5.14 *nothing is emitted from the phosphorus*: If something had been discharged during combustion, it would have shown up as an impurity in the remaining oxygen. But if indeed nothing was emitted, whatever did Lavoisier mean when he spoke of "the vapors emitted" in paragraph 5.12?

It seems clear in retrospect that these "vapors" were the product of combustion in gaseous form (because of the high temperature), which subsequently condensed into "white flakes" as it cooled—leaving pure oxygen as the sole gas in the flask.

5.17 *phosphorus ... is transformed*: The phosphorus that combined with oxygen changed (i) from a cohesive solid to solid flakes, (ii) from being insoluble to highly soluble in water, forming solutions much denser than water, (iii) from a nearly tasteless substance to one whose solution has an "extremely bitter and sharp taste," and (iv) from a combustible to an incombustible substance. The last three of these properties—solubility, sharp taste, and incombustibility—are characteristic of the class of substances called *acids*.

rapidity, and it melts into a liquid much denser than water, and with a much greater specific weight. In the state of phosphorus, before its combustion, it was nearly without taste; by joining with oxygen it takes on an extremely sour and sharp taste. Lastly, it passes from the class of combustibles to the class of incombustible substances and becomes what is called an "acid."

5.18 As we will soon see, this ability to change from a combustible substance to an acid by the addition of oxygen is a property common to a great number of bodies. In accordance with sound logic, it is indispensable to use the same word for all processes which yield analogous results. This is the only way to simplify the study of the sciences, and it would be impossible to remember all the particulars if no attempt were made to classify them. We will therefore call the conversion of phosphorus into an acid, and, in general, [66] the combination of any combustible body with oxygen, "oxygenation."

5.19 We will likewise adopt the expression "to oxygenate," and I will accordingly say that when phosphorus oxygenates, it is converted into an acid.

5.20 Sulphur is also a combustible body; that is to say, it has the property of decomposing air and of removing oxygen from caloric. Experiments just like those I explained in detail above readily establish this. But I should point out that is impossible to obtain as precise results with sulphur as were obtained with phosphorus by means of the same procedure. This is because the combustion of sulphur creates an acid that is difficult to condense, and because

---

In the state of phosphorus: That is, as phosphorus itself, not as part of a compound.

5.20 *a combustible body ... has the property of ... removing oxygen from caloric*: That is, a combustible body has the property of combining with oxygen. Notice this new, implicit definition of "combustion." Far more precise than the casual definition "burning," Lavoisier's characterization identifies the substance-to-substance action that makes combustibility a specifically *chemical* property.

impossible to obtain as precise results with sulphur ... by means of the same procedure: Pay careful attention to the construction of Lavoisier's sentence. He is not saying that precise results cannot be obtained with *sulphur*, only that they cannot be obtained by the *same procedure* as was used for phosphorus. A crucial feature of that procedure was the production of an acid that *condensed into a solid* as it cooled, and so was easy to distinguish from the oxygen gas.

the combustion of sulphur creates an acid that is difficult to condense: If the combustion product remains gaseous, it will be separable from the oxygen only by some additional procedure that was not necessary with phosphorus. (Lavoisier describes methods for separating gases in Part III of the complete Treatise.)

sulphur itself is very hard to burn, and because it is soluble in a number of different gases. But I can assert on the basis of my experiments that when sulphur burns, it absorbs air; that the acid sulphur forms is much heavier than sulphur itself; that the weight of this acid is equal to the sum of the weights of the sulphur and the oxygen it has absorbed; and finally, that this acid is heavy, incombustible, and can combine with water in all proportions. The only uncertainty that remains concerns the quantities of sulphur and oxygen that make up this acid.

[67]

5.21 Charcoal, which should be considered a simple, combustible substance in light of the foregoing, also possesses the property of decomposing oxygen gas and removing the base from caloric. But the acid that results from this combustion does not condense at the temperatures and pressures in which we live. It remains in a gaseous state, and a great quantity of water is required to absorb it. Moreover, this acid has all the properties common to acids, but to a lesser degree, and like other acids, it unites with all bases capable of forming neutral salts.

5.20, cont. *sulphur itself is very hard to burn*: It not difficult to *ignite* sulphur, as Lavoisier's own experiment shows. But if it does not burn *completely*, the experiment will not yield "precise results."

and because [*sulphur*] *is soluble in a number of different gases*: The term "soluble" has a broad range of meanings for Lavoisier, including even *combination* with another substance (compare 1.28, comments). Sulphur does indeed combine with several gases, which would interfere with "precise results." Sulphur also readily passes directly from the solid to the gaseous state—a process that is today called "sublimation" rather than dissolution. But any sulphur lost to sublimation in this experiment would probably be undetectible by weighing.

The only uncertainty ... concerns the quantities of sulphur and oxygen: Lavoisier will reveal in Chapter 6 that there are *two* acids of sulphur, and that one of them readily combines with oxygen to produce the other. Thus he has reason to believe that combustion of sulphur results in a *mixture of two acids*; and this would certainly militate against finding a single, precise ratio of sulphur to oxygen.

5.21 *a great quantity of water is required to absorb it*: Charcoal, like phosphorus and sulphur, forms an acid when it combines with oxygen in the procedure Lavoisier is about to describe. The acid so produced is a gas that is soluble in water—but only to a small degree.

all the properties common to acids: These properties were noted in paragraph 5.17; but Lavoisier here adds one more, the ability to unite with a number of other substances to form "neutral salts." Lavoisier will offer a fuller discussion of such salts in Chapters Sixteen and Seventeen.

5.22 As in the case of phosphorus, the combustion of charcoal can be produced under a glass bell jar A, Plate IV, Figure 3, filled with oxygen gas and inverted in mercury. But since the heat of a hot iron wire, even one that is red-hot, is insufficient to ignite it, a tiny piece of tinder and a speck of phosphorus should be placed on top of the charcoal. The red-hot iron easily ignites the phosphorus, and the fire spreads to the tinder and then to the charcoal.

5.23 The detailed account of this experiment may be found in the memoirs of the Academy.* This report indicates that it requires 72 parts of oxygen by weight to saturate 28 parts of charcoal, and that the resulting aeriform acid has a weight exactly equal to the sum of the weights of the charcoal and oxygen used to produce it. This aeriform acid has been called "fixed air" or "fix'd air" by the first chemists who discovered it. They did not know at the time whether this air was similar to the air of the atmosphere or whether it was another elastic fluid, tainted and corrupted by combustion. But since it is now certain that this aeriform substance is an acid and that it is formed by the oxygenation of a base like all other acids, it is easy to see that the name "fixed air" is inappropriate.

5.24 Once we had burned charcoal in the proper apparatus for determining the quantity of caloric released, Laplace and I found that burning one livre of

[68]

* *Mémoires de l'Académie*, 1781, p. 448.

5.23 *it requires 72 parts of oxygen ... to saturate 28 parts of charcoal*: In keeping with Lavoisier's earlier verb "absorb," the charcoal is said to be *saturated* with oxygen when it has absorbed as much oxygen as it possibly can.

oxygenation of a base: Here again, it is important to keep Lavoisier's use of "base" in mind (1.28, comment).

5.24 *the proper apparatus for determining the quantity of caloric released*: The *ice calorimeter*, depicted in the comment to paragraph 1.43.

charcoal melted 96 livres 6 ounces of ice and that 2 livres 9 ounces, 1 gros, 10 grains of oxygen combine with charcoal in this process, producing 3 livres 9 ounces, 1 gros, 10 grains of acidic gas. This gas weighs 0.695 grains per cubic inch, which means that the total volume of acidic gas produced by the combustion of a livre of charcoal is 34,242 cubic inches.

[69]   5.25 I could provide many additional examples of this kind, and I could present a broad array of facts to show that acids form when substances are oxygenated. But the approach I have determined to take, which consists in proceeding only from what is known to what is unknown and presenting to the reader examples drawn only from what I have already explained, prevents me from presenting certain facts that I will discuss later. Besides, the three examples I have just cited suffice to convey a clear and precise idea of the way acids are formed. It is apparent that oxygen is a principle common to all acids, that it is oxygen that constitutes their acidity, and that it is the nature of the substance acidified that distinguishes acids from one another. In each case, then, the acidifiable base (which de Morveau has named the "radical") must be distinguished from the acidifying principle, that is, oxygen.

5.25 *oxygen is a principle common to all acids*: In the course of the *Treatise*, Lavoisier amasses a wealth of evidence for this general principle. But in the next chapter he will have to acknowledge the anomalous *acid obtained from sea salt*, which he is unable either to produce by combustion or decompose to yield oxygen.

it is oxygen that constitutes their acidity: Lavoisier treats this dictum as though it followed from the previous one. But isn't it possible that oxygen is only an incidental, rather than an essential, ingredient in the acids that contain it? Oxygen, we recall, was a ubiquitous component in common air; but Lavoisier didn't assert that oxygen was what constituted air's ability to support combustion until he showed that *pure* oxygen was *surpassingly* supportive of combustion (3.11). But he cannot demonstrate a similarly essential connection with acidity, since oxygen is not acidic on its own.

the radical: From Latin *radix*, root. Thus de Morveau's term "radical" has a meaning essentially the same as Lavoisier's "base."

Chapter Six

On the Nomenclature of Acids in General, and, in Particular, on Those Extracted from Saltpeter and Sea Salt

6.1 Nothing is easier than establishing a systematic nomenclature for acids on the basis of the principles set down in the last chapter. The word "acid" will be the generic name, and then each acid will be differentiated in language as in nature by the name of its base or radical. In general, we will call whatever is produced from the combustion or oxygenation of phosphorus, sulphur, or charcoal an "acid." We will call the first of these products "phosphoric acid," the second, "sulphuric acid," and the third, "carbonic acid." Likewise, in all such cases we will take the specific name of each acid from the name of the base.

6.2 But there is a remarkable circumstance relating to the oxygenation of combustible bodies, and, speaking generally, to a part of the bodies which transform into acids: they are subject to different degrees of saturation, and the resulting acids, although formed from a combination of the same two substances, have very different properties, properties that depend on a difference of proportion. Phosphoric acid, and especially sulphuric acid, furnish examples of this. If sulphur is combined with only a little oxygen, it becomes, at this first degree of oxygenation, a volatile acid with a pungent odor and very particular properties. A greater proportion of oxygen converts it into a stable, heavy, odorless acid which, when it combines with other substances, yields products very different from the first. Here it seemed that a principle of our nomenclature was found wanting, for it may seem that it would be difficult to derive, from the same acidifiable base, two names which could express these two degrees of saturation without periphrases or circumlocutions.

6.1 *generic ... specific*: As Lavoisier urged in his Preliminary Discourse, substances are to be named according to their genus and species, much as plants and animals are named in the system of Linnaeus; see paragraph 0.24 and comment.

carbonic acid: "*Charbon*" is Lavoisier's word for charcoal, "*l'acide carbonique*" the name of its acid.

6.3 But further reflection and, perhaps especially, necessity revealed to us new resources, such that we came to believe that it might be possible for us to express the different varieties of acids by varying the endings of their names. Stahl chose the name "sulphurous acid" for volatile acid of sulphur. We have kept this name, and have given the name "sulphuric" to the acid of sulphur [72] that is completely saturated with oxygen. And so we will say, using this new language, that when sulphur combines with oxygen, it admits of two degrees of saturation: the first is sulphurous acid, which is volatile and pungent; the second is sulphuric acid, which is odorless and stable. We will adopt the same change in endings for every acid that possesses several degrees of saturation. Thus there will be both phosphorous and phosphoric acid, also acetous and acetic acid, and likewise in other cases.

6.4 This whole part of chemistry would have been extremely simple, and the nomenclature for acids would never have presented difficulties, if the radical or acidifiable base of each acid were known at the time the acid was discovered. Phosphoric acid, for example, was discovered only after the discovery of phosphorus, and the name assigned to it was derived, as a result, from the name of the acidifiable base from which it is formed. But, on the contrary, when an acid was discovered before its base, or rather when, at the time it was discovered, it was not yet known which acidifiable base belonged to it, then the acid and the base were given names which had no relation to each other. [73] This not only burdened the memory with useless names, but even impressed the minds of beginners, as well as the most accomplished chemists, with false ideas that only time and reflection can efface.

6.5 The acid of sulphur will serve as an example. This acid was obtained from vitriol of iron in the early years of chemistry, and so it was called "vitriolic acid." This name was borrowed from the name of the substance from which it was extracted. At that time, no one knew that this acid was the same as the one obtained from sulphur by means of combustion.

6.6 The same may be said of the aeriform acid that was originally given the name "fixed air." No one knew that this acid resulted from the combination

---

6.5 *vitriol of iron*: "Vitriols" were so named because, as colored transparent crystals, they were judged to have a *glass-like* appearance (Latin *vitrum*, glass). The different vitriols were commonly distinguished by color, as blue vitriol, green vitriol, etc. But since the different varieties were so often found in association with specific metal ores, they came to be designated by those metals as well. Thus, blue vitriol was also known as *vitriol of copper*, while green vitriol was *vitriol of iron*.

of charcoal with oxygen. Because of this, it was given any number of different names, none of which conveyed precise ideas. Revising and correcting the names that were used in the older chemical language for these acids could not have been easier. We have changed the name "vitriolic acid" to "sulphuric acid," and "fixed air" to "carbonic acid." But it was not possible for us to take the same approach when it came to naming acids with unknown bases. In [74] these cases we were forced to take the opposite approach, and instead of deriving the name of the acid from its base, we have, on the contrary, named the base according to the name of the acid. This is the approach we took when we named the acid obtained from sea salt or cooking salt. In order to release this acid, simply pour sulphuric acid on sea salt. A lively effervescence arises immediately, pungent white vapors are emitted, and by applying a gentle heat, all of the acid is released. Since the acid occurs naturally in the state of a gas at the degree of temperature and pressure in which we live, special precautions must be taken to retain it.

6.7 The most convenient and simple apparatus for experiments on a small scale consists of a little retort, G, in which some well-dried sea salt has been placed (Plate V, Figure 5). Concentrated sulphuric acid is then poured over the sea salt, and the mouth of the retort is immediately inserted under small vessels or glass bell jars, A, that were previously filled with mercury. As the

Fig. 5.

acid gas is released, it passes into the vessel and makes its way to the top, displacing the mercury. When the release of the gas subsides, gentle heat is applied and gradually increased until no more gas appears. This acid has a [75] great affinity for water: the water absorbs an enormous quantity of it. This may be established by adding a thin layer of water to the glass vessel containing the acid: the acid immediately combines with the water, disappearing entirely. We exploit this affinity in the laboratory and in the arts when we produce acid of sea salt in liquid form by means of the apparatus depicted in

---

6.7 *the acid gas*: That "the acid obtained from sea salt" is indeed an *acid* was so universally accepted, Lavoisier evidently saw no need to list its qualifying properties. He does mention *solubility* later in the paragraph; and the solution has a typically acidic "sharp" taste, which he does not mention. But in comparison with other acids, the "acid from sea salt" is decidedly anomalous, as he will shortly discuss.

Plate IV, Figure 1. This apparatus consists of (1) a retort, A, in which sea salt has been placed, and in which sulphuric acid is poured by means of the short cylindrical opening, H; (2) the spherical glass vessel, CB, set up to receive the small quantity of liquid released; and (3) the series of bottles with two necks L, L′, L″, L‴ that are half-filled with water. This water absorbs the acid gas that is released during distillation. The apparatus is more fully described in the last part of this work.

6.8 Although we have not yet succeeded either in composing or decomposing the acid obtained from sea salt, nevertheless it is undoubtedly created, like all the others, when an acidifiable base is united with oxygen. We have named this unknown base "muriatic base" or "muriatic radical," taking this name, on the model of Bergman and de Morveau, from the Latin word *muria*, formerly a word for sea salt. Thus, although we have not been able to determine the composition of "muriatic acid" exactly, this term, as we will

[76]

6.7, cont. *The apparatus is more fully described...*: Lavoisier's description appears in Part III of the complete Treatise.

6.8 *it is undoubtedly created ... when an acidifiable base is united with oxygen*: What has happened to Lavoisier's noble principle, "one must never infer something that can be ascertained through direct experiment," which he pronounced in paragraph 5.9? For, as he acknowledged at the outset of the present paragraph, his attempts either to *compose the acid from oxygen* or *decompose it to obtain oxygen* have both failed!

It is too easy to fault Lavoisier on this point. We modern readers enjoy the benefit of having been handed an alternative theory, one that singles out *hydrogen* (see Chapter Eight), not oxygen, as the essential ingredient of acids. But should Lavoisier really have abandoned the powerful and beautiful oxygen theory of acids, in the face of all the evidence favoring it, on the strength of a single counter-example? His attempts to demonstrate composition and decomposition "have not *yet* succeeded" [*ne soit encore parvenu*]; that need not mean they will *never* succeed. Moreover, he has a very reasonable explanation for both failures; see the next comment.

use it, will refer to a volatile acid which naturally occurs in the state of a gas at the degree of heat and pressure we typically experience. Great quantities of it combine with water very easily, and its acidifiable radical clings with such strength to oxygen that we have not yet discovered a way to separate them.

6.9 If we someday manage to relate the muriatic radical to some known substance, we will then have to change its name, and give it one analogous to the name of the base whose nature was discovered.

6.10 And something else well worth noting occurs with muriatic acid. Like the acid of sulphur and several others, it admits of different degrees of oxygenation. But adding more oxygen to muriatic acid produces an effect completely opposite to the effect produced in the case of acid of sulphur. The first degree of oxygenation transforms sulphur into a volatile, gaseous acid that only mixes with water in small quantities. With Stahl, we designate this acid by the name "sulphurous acid." A greater measure of oxygen changes sulphurous acid into [77] sulphuric acid, that is, into an acid with more conspicuous acidic properties, an acid that is much more stable, can exist in a gaseous state only at a high temperature, has no odor at all, and unites in great quantities with water. Just the opposite occurs in the case of muriatic acid: the addition of oxygen renders it more volatile, with a more penetrating odor. Less of it can be dissolved in water, and its acidic quality diminishes. We had at first attempted to express these two degrees of saturation, as we had done in the case of the acid of sulphur, by varying the endings. We would have named the acid which is less saturated with oxygen "muriatous acid," and the one more saturated "muriatic acid." But this acid possesses such peculiar attributes that nothing like it is known to chemistry, so we deemed it necessary to make an exception. We decided that a satisfactory name for it would be "oxygenated muriatic acid."

[*the muriatic radical*] *clings with such strength to oxygen that we have not yet discovered a way to separate them*: Such a strong coupling, if true, would obviously explain Lavoisier's failure to "decompose" (6.8) muriatic acid, that is, to release oxygen from it. It would also explain his failure to "compose" (6.8) the acid by oxygenating its radical; since in order to obtain that radical, it would first have to be disengaged from oxygen. But the difficulty has all the appearance of one that must eventually yield to improved techniques. Lavoisier can be excused for not giving up easily.

6.10 *Just the opposite occurs in the case of muriatic acid*: When muriatic acid is oxygenated, its acidic properties become less, not more, pronounced.

oxygenated muriatic acid: For all other acids, the name ending in *-ous* denotes both the lesser degree of oxygenation and the less clearly acidic form, the *-ic* ending signifying the greater degree of oxygenation and the more definitively acidic substance; so those adjectival endings would be misleading if they were applied to the muriatic radical.

Fig. 1.

6.11 There is another acid that we decided to name in the same way as we named muriatic acid, although its base is better known. This is the acid that chemists up until now have referred to as "nitrous acid." This acid is [78] obtained from nitre or saltpeter by a process analogous to the one used to obtain muriatic acid. It is driven from the base to which it is attached in the same manner, by means of sulphuric acid, and, as before, with the apparatus depicted in Plate IV, Figure 1. As the acid passes through the apparatus, part of it condenses in the glass flask while the rest is absorbed by the water in bottles L, L', L'', L''' which becomes green at first, then blue, and then finally yellow, according to the degree of the acid's concentration. During this process, a large quantity of oxygen gas mixed with a small amount of azotic gas is emitted.

6.12 Like all acids, the acid extracted from saltpeter is composed of oxygen united to an acidifiable base, and in fact, the first sound demonstrations of the existence of oxygen in acids were made using this acid. The two principles that constitute it do not have a strong hold on one another, and so they can be easily separated by introducing to oxygen a substance with which it has a greater affinity than it has with the acidifiable base that constitutes the acid of nitre. Experiments like these led to the discovery that nitrous acid

---

6.11 *nitrous acid*: The traditional name "nitrous" was a simple adjective denoting the acid's source (nitre); its *-ous* ending did not have the systematic significance Lavoisier bestows upon *-ous* and *-ic*.

nitre or saltpeter: Known by both names, this substance is found as crystalline deposits on cave walls; it may also be extracted from guano deposits, using water.

6.12 *the first sound demonstrations of the existence of oxygen in acids were made using this acid*: The demonstrations were conducted by Lavoisier himself in 1776. He heated the acid with mercury; at the end of a lengthy process, the mercury was recovered intact along with a mixture of gases, one of them oxygen.

contains azote, the base of noxious gas, which is its acidifiable base. Azote is therefore the true nitric radical, or, in other words, the acid of nitre is a true azotic acid. So it seemed that in order to remain consistent with ourselves and with our principles, we would have had to adopt one of these two ways of expressing ourselves.

[79]

6.13 However, other considerations deterred us. First of all, it seemed difficult to us to change the names "nitre" and "saltpeter," which are used throughout the arts, in society, and in chemistry. Furthermore, we thought that we should not call azote the "nitric radical," because azote is also the base of volatile alkali (ammoniac), as Berthollet has discovered. We will therefore continue to refer to the base of the non-respirable part of atmospheric air as "azote," which is at the same time the nitric radical and the ammoniac radical. We will also retain the words "nitrous" and "nitric" for the acid drawn from nitre or

---

nitrous acid contains azote ... which is its acidifiable base: That nitrous acid "contains azote" is no more than what Lavoisier acknowledged earlier, in paragraph 4.10. But that was before he determined that every acid is distinguished by a corresponding acidifiable base (5.25). Now that the general constitution of acids is established, he is able to report that azote is not merely *a* constituent; it is *the* specific acidifiable base of nitrous acid.

it seemed that ... we would have had to adopt one of these two ways of expressing ourselves: The two alternatives are (i) "azote is the true nitric radical" and (ii) "the acid of nitre is azotic acid." The first locution would relinquish the name azote. Lavoisier might have taken that path in Chapter Four: instead of styling the non-respirable part of air *azote*, he might have called it the nitric radical, reflecting its status as the base of the acid of nitre. The second locution would abandon the name nitre, replacing it with a term that derives from *azote*, and so renaming "acid of nitre" as "azotic acid." But "other considerations deterred" him from doing either.

6.13 *it seemed difficult to us to change the names...*: He did, indeed, state in the Preliminary Discourse that "it was our aim to retain all common or customary names" (0.22).

Furthermore, we thought that we should not call azote the "nitric radical," because azote is also the base of volatile alkali...: Lavoisier's implicit argument here is that we have no way to decide which property is more fundamental—the ability to form nitre or the ability to form volatile alkali. His reason for rejecting the term "nitric radical" here is the same as his reason for declining "nitrogenic radical" in paragraph 4.10.

We will also retain the words "nitrous" and "nitric" for the acid drawn from nitre: Nevertheless, he will hold out for one innovation, the systematic use of *-ous* and *-ic* to indicate degrees of oxygenation. Lavoisier will shortly identify two kinds of "acid of nitre"; but for many years the traditional name "nitrous acid" was used indiscriminately for both.

saltpeter. Several chemists of great authority have disapproved of the way we have stooped to these older names. They would have preferred us to have focused all of our efforts on the perfection of nomenclature, to have reconstructed the edifice of chemical language from top to bottom without troubling ourselves to make connections with former usages, the memory of which time will gradually obliterate. And so we have found ourselves subjected to the criticisms and complaints of the two opposing sides simultaneously.

[80] 6.14 The acid of nitre can occur in many different states depending on its degree of oxygenation, that is to say, on the proportions of azote and oxygen which compose it. A first degree of oxygenation of azote yields a particular gas, which we will continue to call "nitrous gas." It is composed of approximately 2 parts oxygen and one part azote by weight, and in this state it does not mix with water. The azote in this gas, far from being saturated with oxygen, retains a great affinity for it, and attracts it with such vigor that it draws oxygen away from any atmospheric air which comes in contact with it. In fact, we now combine nitrous gas with atmospheric air in order to determine the quantity of oxygen the air contains and to gauge its wholesomeness. This addition of oxygen changes nitrous gas into a powerful acid that has a great affinity for water, and which is itself capable of different degrees of

---

6.13, cont. *Several chemists of great authority have disapproved*: They were evidently not as hesitant about the ambivalent choice of names as Lavoisier was. As early as 1791 the distinguished chemist Jean-Antoine Chaptal proposed replacing the name "azote" by "nitrogen." That name, he wrote, "is deduced from the characteristic and exclusive property of this gas, which forms the radical of nitric acid." Modern chemistry has unswervingly adopted this path.

6.14 *A first degree of oxygenation of azote yields ... nitrous gas*: Notice that this lowest degree of oxygenation—two parts oxygen to one of azote—is not even called *acid*.

In fact, we now combine nitrous gas with atmospheric air in order to determine the quantity of oxygen the air contains: This was the "nitrous air test" developed by Joseph Priestley. He would mix a sample of common air with nitrous gas confined over mercury. The nitrous gas would snatch oxygen from the sample, and the volume of the mixture would decrease accordingly. The more oxygen in the sample, the greater the decrease—and the more "wholesome" was the sample judged to be.

addition of oxygen changes nitrous gas into a powerful acid: This represents a second degree of oxidation of azote, and the lowest degree that produces an acid. Lavoisier will call it "nitrous acid" later in the paragraph.

oxygenation. If the proportion of oxygen to azote is below three parts to one, the acid is red and fumes profusely. In this state we call it "nitrous acid." This acid releases nitrous gas when it is gently heated. Four parts of oxygen to one part azote produces a white, colorless acid, which is more stable when exposed to heat than nitrous acid, and less odorous, and whose two constitutive principles are more strongly united to one another. In accordance with the principles outlined above, we have given it the name "nitric acid." [81]

6.15 Thus, of these two acids, nitric acid is the acid of nitre with the greater proportion of oxygen, and nitrous acid is the acid of nitre with the greater proportion of azote (or, what amounts to the same thing, with the greater proportion of nitrous gas). Finally, nitrous gas is azote that is not sufficiently saturated with oxygen to possess the properties of acids. Later we will call this an "oxide."

If the proportion of oxygen to azote is below three parts to one, ... we call it "nitrous acid": The proportion Lavoisier cites is by volume, not weight.

Four parts oxygen to one part azote produces ... "nitric acid": This represents the third degree of oxygenation of azote, and the second that produces an acid.

6.15 *...not sufficiently saturated with oxygen to possess the properties of acids. Later we will call this an "oxide"*: Oxides will be the principal topic of the next chapter.

Chapter Seven

On the Decomposition of Oxygen Gas by Metals,
and on the Formation of Metallic Oxides

7.1 When metallic substances are heated to a certain temperature, oxygen has more affinity for them than for caloric. Because of this, all metallic substances, except gold, silver and platinum, have the property of decomposing oxygen gas, taking hold of its base, and disengaging caloric. We have already seen above how this decomposition of air by mercury and iron takes place, and we have observed that the decomposition by mercury can only be considered a slow combustion, whereas decomposition by iron, on the contrary, is very quick and is accompanied by a brilliant flame. If a certain degree of heat is required for these procedures, this is only in order to separate the particles of the metal from one another so as to diminish their affinity of aggregation, or, what is the same thing, the attraction that they exert on one another.

7.2 Metallic substances, during calcination, increase in weight in proportion to the oxygen that they absorb, and at the same time they lose their metallic sheen and are reduced to an earthy powder. Metals in this state ought not to be considered completely saturated with oxygen, since the effect of metals on oxygen is balanced by the force of attraction that caloric exerts on

---

7.1 *we have observed that the decomposition [of oxygen] by mercury can only be considered a slow combustion*: Here, as previously, "decomposition" of a gas means releasing the gas's base from caloric. Mercury accomplishes this by combining with the base. Lavoisier has not explicitly affirmed that calcination of mercury is "slow combustion"; but the interpretation is inescapable, since calcination and combustion are both processes of *union with oxygen*.

7.2 *Metals in this state...*: That is, the state of having combined with oxygen.

...ought not to be considered completely saturated with oxygen: Why is Lavoisier so quick to caution readers against this supposition? Possibly because when he calcined mercury and iron in Chapter Three, nothing happened to suggest that these metals were capable of combining with a greater quantity of oxygen than they actually did. Although it might then be natural to entertain the possibility that these metals were capable of only a single state of oxidation—which would necessarily be a state of saturation—it would be rash to assume it.

the effect of metals on oxygen is opposed by the ... attraction that caloric exerts on oxygen: Lavoisier previously called attention to such a rivalry in paragraph 3.14.

oxygen. Thus, in the calcination of metals, oxygen is actually subject to two forces, the force exerted by the caloric, and the force exerted by the metal, and it only tends to unite with the metal because of the difference between these two forces, that is, as a result of the extent to which the force exerted by the metal exceeds the force exerted by the caloric. And in general this excess in not very significant. So metallic substances, unlike sulphur, phosphorus, and charcoal, are by no means converted into acids when they oxygenate in the air or in oxygen gas. They become, rather, intermediary substances: they begin to approach a saline state without acquiring every saline property.

7.3 Earlier chemists used the term "calx" to refer not only to any metal that had been brought to this state, but to any substance that did not melt after long exposure to fire. In this way, they used "calx" as a generic term, under which they jumbled together limestone with metals; that is to say, they jumbled together a substance which, in calcination, is converted from a neutral salt to an earthy alkali and loses half its weight, with sub-stances which, in calcination, gain sometimes more than half their weight and become almost acids. It would have been contrary to our principles to [84]

So metallic substances ... are by no means converted into acids when they oxygenate: He seems to be of the opinion that metals simply cannot exert a strong enough hold upon the oxygen they combine with to give them acidic properties.

they begin to approach a saline state: A "saline" state (from Latin *sal*, salt) is one that resembles, or contains, salt. In Chapters Sixteen and Seventeen Lavoisier will note that oxidized metals become salts *when acted upon by acids*. Lavoisier terms them the "salifiable bases" of their respective salts; and he labels acids as "the true salifying principles." He also acknowledges *earths* and *alkalis* as two other classes of salifiable bases. We can perhaps begin to see how the term "base" could gradually develop its modern connotation as the counterpart to "acid."

7.3 *calx*: Lavoisier has been scrupulous in using this term exclusively to denote a metal that has been heated sufficiently to combine with oxygen and so gain weight. But since others have used "calx" so loosely, he will in paragraph 7.4 adopt the term "oxide" instead.

a substance which, in calcination, is converted ... to an earthy alkali: The "earthy alkali" he has in mind here is commonly called *quicklime*. Previous chemists were wont to call quicklime a "calx" because it was produced by *calcination* of limestone.

substances which, in calcination, ... become almost acids: When substances are calcined they unite with oxygen; and oxygenation produces either an acid (5.25) or a state that is on the way to becoming an acid (6.15). So by calling all products of calcination "calces," earlier chemists lumped alkaline and acidic substances into the same class, even though their properties are very different.

classify under the same name substances so different and, above all, to keep a name for metals so likely to give rise to false ideas. Because of this we have banned the term "metallic calces" and have substituted in its place the term "oxides" from the Greek ὀξύς.

7.4 From this it is clear how fruitful and expressive is the language we have adopted: a first degree of oxygenation forms oxides; a second degree forms acids with names ending in "ous," such as nitrous acid, and sulphurous acid; a third degree forms acids with names ending in "ic" such as nitric and sulphuric acids, and, finally, we can express a fourth degree of oxygenation of these substances by adding the qualifying adjective "oxygenated," as we noted in the case of oxygenated muriatic acid.

[85] 7.5 And by the term "oxides" we refer not only to those oxygen compounds involving metals. For we saw no reason not to use this term to express the first degree of oxygenation of all substances which [when so oxygenated] do not become acids but, rather, approach a saline state. Thus we will call sulphur, when it first begins to burn and becomes soft, "oxide of sulphur," and we will refer to the yellow substance that remains after phosphorus burns "oxide of phosphorus."

7.6 We will also call nitrous gas, which is the first degree of oxygenation of azote, "oxide of azote." And, finally, there will be oxides in the animal and vegetable kingdoms; and I will show in what follows how this new language illuminates every operation of art and nature.

7.7 As we have seen, almost every metallic oxide has its own particular color. And these colors vary not only from metal to metal but even among the different degrees of oxygenation of the same metal. We thus found ourselves obliged to add to each oxide two epithets, one indicating the metal oxidized, and one indicating its color. Thus we will use the terms "black iron oxide," "red iron oxide," and "yellow iron oxide" to refer to the substances previously known as "martial ethiops," "colcothar," and "iron rust" (or "ochre") respectively.

[86] 7.8 In the same way we will speak of "grey lead oxide," "yellow lead oxide," and "red lead oxide," and these expressions will designate the substances formerly known as "lead ash," "massicot" and "minium" respectively.

7.3 *we have banned the term "metallic calces" and have substituted ... the term "oxides"*: Even if the common usage of "calx" had not robbed the term of precision, it would still fail to express the *composition* of its referent (0.28), conveying instead merely the method for producing it. "Oxide" indicates combination with oxygen— an attribute that penetrates far more deeply into the nature of the substance.

7.9 The new names will sometimes be rather long, especially when it is necessary to express whether the metal has been oxidized in the air, by detonation with nitre, or through the action of acids, but at least they will always be just and will give rise to the precise idea of the object which corresponds to them.

7.10 The tables appended to this treatise will make all of this clearer.

7.9 *detonation with nitre*: A *detonation* is an explosion accompanied by a sudden loud report; from Latin *tonare*, thunder. In Part II of this work Lavoisier writes: "One can also oxygenate combustible substances by combining them ... with nitrate of potash [nitre]. At a certain degree of heat, the oxygen leaves the nitrate ... in order to combine with the combustible body. However, these sorts of oxygenation must be attempted with extreme precautions and with very small quantities. Oxygen combines with nitrates with [an] immense quantity of caloric [which] becomes suddenly free at the moment of its combination with the combustible bodies, resulting in dreadful detonations which nothing can withstand."

through the action of acids: Acids typically act slowly on most metals, although zinc and some others dissolve quickly in acids, producing an effusion of bubbles ("effervescence"). In contrast, acids act rapidly, without effervescence, on oxidized metals. In both cases, a salt is produced. So when Lavoisier speaks of a metal being *oxidized* by action of an acid, he means the production of a *salt*, not a *metal oxide*: if an oxide were produced, it would immediately react with the acid to form the salt (7.2, comments). Lavoisier regards the metal constituent of a salt as "oxidized" insofar as both it and oxygen, being constituents of the same substance, are in combination with one another. Lavoisier will discuss the salification of metals, identifying the stages of that process, in Chapter Seventeen.

7.10 *The tables appended to this treatise will make all of this clearer*: The tables follow Part III of the complete Treatise.

Chapter Eight
On the Radical Principle of Water, and on the Decomposition of Water by Charcoal and by Iron

8.1 Until very recently, water had been considered a simple substance, and our predecessors had had no qualms calling it an element. They never doubted that it was an elementary substance since they had not yet reached the point of decomposing it, or rather, because the decompositions of water which took place before their eyes escaped their notice. But we shall soon see that water is no longer an element for us. I will not give an account of the history of this discovery here, which is very recent and still controversial.*

8.2 I will content myself with reporting the main proofs of the decomposition and recomposition of water. I will go so far as to say that when these proofs are weighed without prejudice they will be found to be conclusive.

First experiment.
Set-up

8.3 Take a glass tube, EF, with a diameter of 8 to 12 lines, and place it over a furnace at a slight incline from E to F (Plate VII, Figure 11). Attach a small glass retort, A, filled with common distilled water to the upper end of the tube, E, and to the lower end, F, attach a spiral tube, SS′, which has been attached at its other end, S′, to one mouth of a double-mouthed flask, H. And, finally, attach to H's other mouth, a curved glass tube, KK, designed to conduct aeriform fluids, or gases, into a device appropriate for determining their quantity and quality.

8.4 In order to ensure that the experiment succeeds, the tube EF should be made from well-annealed green glass which is difficult to melt. Coat the outside with a lute of clay mixed with a cement made of pulverized stoneware pottery, and in case the glass should bend as it melts, it should be supported in the middle with an iron bar across the furnace. Porcelain pipes are preferable to glass pipes, but it is difficult to obtain porcelain pipes which are not

porous. These pipes are almost always found to have some holes that allow air or vapors to pass through.

---

* For this, see *Mémoires de l'Académie des Sciences*, 1781.

Fig . 11 .

8.5 When everything has been set up as indicated, light a fire in the furnace, EFCD, and stoke it in such a way as to make the tube EF glow red without melting it. At the same time, light a fire in the furnace, VVXX, so as to keep the water in retort A boiling.

Result

8.6 As the water in the retort boils and evaporates, it fills the interior of the tube EF, and expels the common air which leaves by the tube KK. Aqueous gas then condenses as it cools in the spiral tube SS' and water falls, drop by drop, into the flask H.

8.7 When this procedure is continued until all the water in the retort is evaporated and the vessels are allowed to drain completely, flask H is found to contain a quantity of water exactly equal to that which was in the retort A, without any gas having been released. We have thus ensured that this procedure is nothing but simple, ordinary distillation, and the outcome is absolutely the same as it would have been if the water had not been brought to an incandescent state as it passed through the intermediary tube EF.   [90]

Second experiment.
Set-up

8.8 Set everything up as in the previous experiment, with one difference: place 28 grains of charcoal in the tube EF. The charcoal should be crushed into

---

8.6 *aqueous gas*: That is, water in the gaseous state (4.7). It condenses into liquid water.

8.7 *an incandescent state*: Here, as also in the following paragraph, "incandescent" means *intensely hot*, rather than its more literal signification, "glowing." Compare Lavoisier's expression "red heat" in paragraph 8.14, where the *container*, not the substance being heated, becomes incandescent.

middle-sized pieces and exposed beforehand to incandescent heat in closed vessels. Then boil water in retort A until it completely evaporates as in the preceding experiment.

Result

8.9 Water from the retort is distilled in this experiment as in the preceding experiment. It condenses in the spiral tube and drips drop by drop into flask H. But at the same time it releases a considerable quantity of gas, which escapes by tube KK, and is collected in an appropriate receptacle.

8.10 At the end of the process, only a few specks of ash remain in tube EF: the 28 grains of charcoal have completely disappeared.

[91]   8.11 When the gases released are examined carefully, they are found to weigh a total of $113\frac{7}{10}$ grains.* There are two types of gas: 144 cubic inches of carbonic acid gas, weighing 100 grains, and 380 cubic inches of an extremely light gas which weighs $13\frac{7}{10}$ grains and ignites in the presence of a flame when it is in contact with the air. If the weight of the water that has passed into the flask is checked, it is found to have diminished by $85\frac{7}{10}$ grains.

---

* The third part of this work includes a detailed account of the method for separating different types of gas and for weighing them.

8.8 *exposed beforehand to incandescent heat in closed vessels*: This process drives off any volatile impurities. Such a degree of heat might trigger calcination or combustion of the charcoal (5.21–5.22), although the limited air supply in "closed vessels" helps limit the extent to which this can occur. But since any combination with oxygen will diminish the supply of solid charcoal, the heating is to take place "beforehand"—that is, before weighing.

8.11 *two types of gas*: From Chapter Five, we know that the 100 grains of carbonic acid gas represents the combination of 28 grams charcoal and 72 grams oxygen (5.23).

extremely light: Since 380 cubic inches weigh $13\frac{7}{10}$ grains, one cubic inch weighs .036 grains, about one-fourteenth the weight of a cubic inch of oxygen (4.8).

8.12 Thus, in this experiment, $85\frac{7}{10}$ grains of water plus 28 grains of charcoal has formed 100 grains of carbonic acid plus $13\frac{7}{10}$ grains of a certain flammable gas. But I have shown above, that in order to form 100 grains of carbonic acid gas, it is necessary to combine 72 grains of oxygen with 28 grains of charcoal. Thus the 28 grains of charcoal placed in the tube removed 72 grains of oxygen from the water, which means that $85\frac{7}{10}$ grains of water is composed of 72 grains of oxygen and $13\frac{7}{10}$ grains of flammable gas. We shall soon see that we cannot imagine that this gas was released from the charcoal, and that it is consequently a product of water.

8.13 In my account of this experiment, I have suppressed certain details [92] that would have only tended to complicate it and obscure readers' ideas. The flammable gas, for example, dissolves a little of the charcoal, and this factor increases its weight and on the contrary diminishes the weight of the carbonic acid. The resulting differences in the quantities are not very considerable, but I thought I should recalculate them and present the experiment in its complete simplicity, as if no charcoal had dissolved, in order to present the experiment in as simple a form as possible. Moreover, if any suspicions obscure the truth of these conclusions, they will be dispelled by the other experiments I will relate in support of them.

Third experiment.
Set-up

8.14 Arrange all of the equipment in the same way as in the preceding experiment, except instead of 28 grains of charcoal, place 274 grains of soft iron strips shaped into spirals in the tube, EF. Bring the tube to a red heat as in the preceding experiment, and light the fire under the retort, A, keeping [93]

8.12 *I have shown above*: In paragraph (5.23).

85.7 grains of water is composed of 72 grains of oxygen and $13\frac{7}{10}$ grains of a certain flammable gas: He thus obtains 5.25 to 1 as the ratio, by weight, of oxygen to flammable gas.

We shall soon see that we should not imagine that this gas was released from the charcoal: This will become apparent in the Third Experiment, especially paragraphs 8.15 and 8.16.

8.13 *The flammable gas ... dissolves a little of the charcoal*: Here, again, "dissolves" probably does not refer to *sublimation* (5.20, comment), since that would not diminish charcoal's availability to form carbonic acid. But at temperatures well within reach of Lavoisier's furnace EFCD, a small amount of charcoal might *combine* with the flammable gas to form another gas that would later be called "methane."

recalculate ... as if no charcoal had dissolved: Lavoisier does not tell us what method he uses to make this correction.

the water at a constant boil until it has completely evaporated and all of it has passed through the tube EF and condensed in flask H.

Result

8.15 No carbonic acid gas is released in this experiment. The only gas released is flammable and 13 times lighter than atmospheric air. Its total weight is 15 grains, and it has a volume of around 416 cubic inches. If the initial quantity of the water used in this experiment is compared to the water remaining in flask H, it is found to have lost 100 grains. On the other hand, the iron contained in tube EF, which initially weighed 274 grains, gains 85 grains, and its volume increases considerably. This iron is hardly attracted to the magnet anymore, and it dissolves in acids without effervescence. In a word, it becomes a black oxide, precisely as if it had burned in oxygen gas.

Reflections

[94]   8.16 The result of this experiment is true oxidation of iron by water, oxidation exactly like that which occurs in the air with the aid of heat. 100 grains of water has been decomposed, 85 grains of oxygen has been united to the iron so as to constitute a black oxide, and 15 grains of a certain flammable gas has been released. Thus, water is composed of oxygen and the base of a flammable gas in the ratio of 85 parts to 15.

8.17 Thus, water contains not only oxygen, which is one of its principles and which it shares in common with many other substances, but another principle which is proper to it, its constitutive radical, and which we were obliged to name. No name seemed more appropriate than "hydrogen," that

---

8.15 *The only gas released is flammable and 13 times lighter than atmospheric air*: This is clearly the same gas as that produced in the second experiment.

the iron ... gains 85 grains: Since its initial weight was 274 grains, it increased in weight by 31 percent—in decent agreement with the 35 percent weight increase when he oxidized iron in Chapter Three (3.24).

it dissolves in acids without effervescence: As noted in the comment to paragraph 7.10, this is characteristic of metal oxides.

In a word, it becomes a black oxide, ... as if it had burned in oxygen gas: This black oxide of iron is identical to the *martial ethiops* produced earlier (3.24). The iron that formed it must, therefore, have united with oxygen and only oxygen; and the only possible source for that oxygen is the water. Then the "flammable gas" must represent a hitherto unknown constituent of water that remains behind when the oxygen is removed.

8.16 *water is composed of oxygen and the base of a flammable gas in the ratio of 85 parts to 15*: That is, a ratio of 5.7 to 1—a value reasonably consistent with the ratio of weights obtained in the previous experiment, 5.25 to 1 (8.12, comment).

is to say the "generating principle of water" from ὕδωρ ("water") and γείνομαι ("I generate"). We will call "hydrogen gas" the combination of this principle with caloric, and the word "hydrogen" will only express the base of this same gas, the radical of water.*

8.18 Here, then, is a new kind of combustible body, that is to say, a body with enough affinity for oxygen to remove it from caloric and to decompose air or oxygen gas. This combustible body has, itself, so much affinity for caloric, that unless it is engaged in a compound, it is always in an aeriform or gaseous state at the ordinary pressures and temperatures in which we live. In [95] this gaseous state it is about 13 times lighter than atmospheric air, water does not absorb it (though it can dissolve a small quantity of water), and it is not fit for the respiration of animals.

8.19 Since the property of burning and flaring up is, for this gas and for all other combustible gases, nothing but the property of decomposing the air and removing oxygen from caloric, it is clear that the gas cannot burn unless it is in contact with air or oxygen gas. Thus, when a bottle is filled with this gas and the gas is lit, it burns gently at the mouth of the bottle and only afterwards inside as the air outside makes its way in. But the combustion is successive and slow: it only occurs at the point of contact between the two gases. It is not the same when one mixes the two airs together before lighting them. If, for example, a burning body (such as a piece of paper or a candle) is brought close to the mouth of a narrow-neck bottle filled with one part oxygen gas and then two parts hydrogen gas, the combustion of the two gases occurs instantaneously [96] and with a strong explosion. This experiment should only be made with a very strong green glass bottle with a volume of no more than a pint and which has been covered with a rag, since otherwise disastrous mishaps can occur: if the bottle ruptures, fragments of glass can shoot out to great distances.

8.20 If all that I have set out concerning the decomposition of water is accurate and true—if this substance is really composed, as I have attempted to establish, of a principle (hydrogen) proper to it, combined with oxygen— it should be the case that by uniting these two principles, one ought to be

* I have been bitterly criticized for this expression, "hydrogen," because it has been claimed that that term signifies not "the generating principle of water" but "the offspring of water." But it hardly matters, since either interpretation is appropriate. The experiments described in this chapter prove that water, when decomposed, "gives birth" to hydrogen, and moreover hydrogen "gives birth" to water when it combines with oxygen. One could just as well say that hydrogen "generates" water as that water "generates" hydrogen. [Lavoisier added this note in the 1793 edition. —Tr.]

8.19 *the combustion is successive*: It takes place gradually, not all at once as in the *explosion* mentioned later in the paragraph.

able to create water again. The following experiment demonstrates that this is exactly what happens.

Fourth experiment: Recomposition of water.
Set-up

8.21 Take a spherical crystal flask, with a large opening, and a volume of around 30 pints, and attach a copper plate, BC, to the flask with putty (Plate IV, Figure 5). Four tubes should pass through holes pierced in the plate.

[97]   8.22 The first tube, Hh, is designed to be attached at one end, h, to an air pump which can be used to produce a vacuum in the flask. A second tube, gg, is attached at one end, MM, to a reservoir of oxygen gas, and is designed to conduct the gas into the spherical flask. The third, dDd' is attached at one end, dNN, to a reservoir of hydrogen gas. There should be such a small opening at the end of this tube, d', that a very fine needle can hardly pass through it. The hydrogen gas in the reservoir must escape through this tiny opening, and in order to ensure that it escapes quickly enough, it ought to be subjected to a pressure of 1 or 2 inches of water. Finally, the fourth hole in plate BC is fitted with a glass tube which is attached to the plate with putty, and through which passes a metal wire, GL. A little ball is designed to be attached to the end L of the wire which makes it possible to draw off electric sparks from L to d' and, as will soon be apparent, ignite the hydrogen gas. The metal wire is free to move in the tube so that the ball can be separated from the end d' of the tube Dd'. The three tubes dDd', gg, Hh are all fitted with stopcocks.

8.23 The hydrogen and oxygen gases should arrive very dry from their respective tubes that bring them to the flask A, and they should have as much water as possible removed from them. Therefore, make them pass through tubes MM and NN about an inch in diameter, which are filled with exceptionally deliquescent salt—salt, that is, which sucks up the moisture of the air voraciously, such as acetate of potash, or muriate or nitrate of lime. For the purposes of this experiment, these salts ought to be ground to a coarse powder, so that they do not cake, and so that the gas can pass easily through the interstices between the particles. We will examine the composition of these salts in the second part of this work.

8.24 Make sure in advance that there is a sufficient supply of very pure oxygen gas, and in order to make certain that this gas does not contain any carbonic acid, allow the gas to remain in contact with potash that has been dissolved in water and from which all carbonic acid has been removed by means of lime. Details for obtaining this alkali are provided below.

8.25 The same care should be taken in preparing a double portion of hydrogen gas. The surest method for obtaining the gas unmixed consists in extracting it in the decomposition of water with very pliant and very pure iron.

8.26 When the two gases have been thus prepared, attach the air pump to the tube Hh, and make a vacuum in the large spherical flask A. Introduce one of the two gases, preferably the oxygen first, by means of the tube gg. Then, using a certain degree of pressure, introduce the hydrogen through the

---

8.23 *the second part of this work*: Not included in the this volume, Part II of the Treatise is titled "On the Compounds of Acids with Salifiable Bases, and on the Formation of Neutral Salts."

8.24 *Details for obtaining this alkali*: That is, for obtaining potash. The procedure will be found in Chapter Sixteen.

8.25 *a double portion of hydrogen*: Lavoisier means this merely as a practical hint. Having performed the experiment previously, he knows that unless he starts with a double volume, he will run out of hydrogen before he consumes the oxygen. But in 1804, Joseph-Louis Gay-Lussac will discover that the 2:1 volume ratio for producing water is *exact*. That finding, together with the now-accepted densities of hydrogen and oxygen, implies that the ratio of oxygen to hydrogen in water *by weight* is 8 to 1, rather than the ratios Lavoisier obtains in paragraphs 8.12, 8.16, and 8.28.

extracting it from the decomposition of water with very pliant and very pure iron: Ideally, the Third Experiment produces hydrogen, unmixed with carbonic acid gas (8.15); but that acid will be formed if the iron used contains carbon (Lavoisier's "charcoal"), since any carbon will oxidize. *Pliant* iron is soft and bendable—a sign that it contains little or no carbon. (Iron with an admixture of carbon up to about 2% is *steel*, which is not pliant but rigid.) Compare paragraph 3.28.

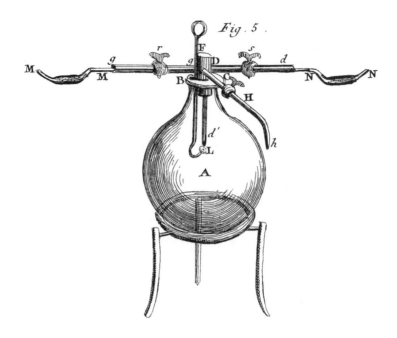

tube dDd', whose extremity d' terminates in a point. Finally, ignite the gas with the help of an electric spark. By letting in a little of both gases into the flask in this way, it is possible to continue the combustion for a long time. I have provided elsewhere a description of the equipment that I used for this experiment and have explained how it is possible to measure with rigorous precision the amount of the two gases consumed. See the third part of this work.

Result

8.27 As the combustion proceeds, water is deposited on the interior walls

---

8.26 *the equipment that I used*: Above all, the appliance called the *gasometer*, illustrated here. This gargantuan apparatus was capable of delivering a measured quantity of gas at a controlled pressure. In this experiment it provided a constant flow of gas to the combustion flask depicted in Figure 5.

See the third part of this work: Part III of the complete *Treatise* contains Lavoisier's fascinating review of the chemical instruments and operations employed in his investigations.

of the flask. The quantity of this water increases bit by bit, and it collects in large droplets which run together at the bottom of the vessel.

8.28 By weighing the flask before and after the procedure, it is easy to determine the amount of water that forms. Thus the experiment allows us to double-check our work: on the one hand we have the weight of the gas employed, on the other the weight of the water formed, and these two quantities ought to be equal. By means of an experiment like this one, Meusnier and I found that it takes 85 parts by weight of oxygen and 15 parts by weight of hydrogen to create 100 parts of water. This experiment, which has still not been published, was performed in the presence of a large committee from the Academy. We attended to the experiment meticulously and have reason to believe that the ratio is exact within no more than 2 percent. [100]

8.29 And so, whether from the perspective of decomposition or recomposition, we can consider the following facts to be as invariable and as well-established as any facts can be in chemistry and natural philosophy: that water is not a simple substance; that it is composed of two principles, oxygen and hydrogen; and that these two principles, when separated from one another, have such an affinity for caloric that they can only exist in the form of gases at the degree of temperature and pressure in which we live.

8.30 This phenomenon of decomposition and recomposition of water occurs constantly before our eyes at atmospheric temperature due to the compounded affinities of the substances involved. At least to some extent, [101]

8.28 *these two quantities ought to be equal*: That is, the total weight of the two gases, hydrogen and oxygen, should equal the weight of water produced. Here again, as in paragraph 5.7, Lavoisier is tacitly employing the principle of constancy of weight.

85 parts by weight of oxygen and 15 parts by weight of hydrogen: Exactly the same ratio, this time for *composition* of water, as he found by *decomposition* in the Third Experiment. Notwithstanding Lavoisier's confident estimate of accuracy within 2 percent, it is very far from the 8 to 1 ratio which later investigators will determine and which is still the accepted value today (8.25, comment). But no one disparages Lavoisier's luminous demonstration that water is not an element.

8.29 *water is not a simple substance*: That water is an element was of course a tenet of Aristotelian and other ancient schools of natural philosophy. But the true importance of Lavoisier's finding has little to do with confutation of the ancients. Rather it illustrates what Lavoisier has already expressed in his Preliminary Discourse: that an element or simple substance is nothing more than one that we have so far been unable to decompose—and that even substances long accepted as "elements" may at some point prove to be compound in nature.

phenomena such as the fermentation of spirits, putrefaction, and even the growth of plants depend on the decomposition of water, as will soon be apparent. So it is quite extraordinary that this decomposition escaped the watchful eyes of natural philosophers and chemists until recently, and from this we ought to conclude that in science, as in ethics, it is difficult to overcome deeply-entrenched prejudices and to take any path not already well worn.

8.31 I will conclude this chapter with an experiment much less decisive than those I have already described, but which, however, seems to me to have made more of an impression than any other on many people. If one burns a livre or 16 ounces of spirit of wine, i.e., alcohol, in a device designed to collect all the water released during the combustion, one obtains 17 or 18 ounces.* Now no material can increase in weight during an experiment, so it must be that another substance affixes itself to the spirit of wine during the combustion. And I have shown that this other substance is the base of the air, or oxygen. Spirit of wine thus contains one of the principles of water (hydrogen) and atmospheric air provides the other (oxygen). This is a new proof that water is a composite substance.

[102]

* See the description of this device in the third part of this work. .

8.30 *phenomena such as the fermentation of spirits ... depend on the decomposition of water*: In Chapter Thirteen Lavoisier will moderate his assertion that water decomposes in vinous fermentation (winemaking).

8.31 *a device designed to collect all the water released*: The apparatus, which is discussed in Part III of the complete Treatise, is pictured here. Alcohol (or other combustible liquid) in the reservoir L, at the lower left, supplies a lamp at the base of the vertical chimney. When the liquid burns, its gaseous products are routed through the spiral "worm." If those gases include water (as they invariably do), that water condenses and collects in the flask P at the lower right.

Fig. 5 .

Chapter Nine

On the Quantity of Caloric Released from Different Types of Combustion

9.1 We have seen that when any sort of combustion is produced in a hollow sphere of ice, provided with the means of maintaining the air at a temperature of zero degrees, the quantity of melted ice in the interior of the sphere is a measure, if not absolute, at least relative, of the quantity of caloric released. Laplace and I gave a description of the apparatus we used in these kinds of experiments.* Having attempted to ascertain the quantity of ice melted in the combustion of three of the four simple combustible substances, namely, phosphorus, carbon, and hydrogen, we obtained the following results:

For the combustion of a livre of phosphorus:
> 100 livres of ice.

For the combustion of a livre of carbon:
> 96 livres, 8 ounces of ice.

For the combustion of a livre of hydrogen gas:
> 295 livres, 9 ounces, $3\frac{1}{2}$ gros of ice.

9.2 As the substance formed from the combustion of phosphorus is a solid acid, it is likely that very little caloric remains in this acid. Consequently this combustion gives us a way of finding out, at least for the most part, the amount of caloric contained in oxygen gas. But even if we should suppose

---

* See *Mémoires de l'Académie des Sciences*, 1780, page 355. Also see the third part of this work.

9.1 *a measure, if not absolute, at least relative, of the quantity of caloric*: An "absolute" measure of caloric would be expressed in units of caloric, if such units could be obtained. In their absence, Lavoisier expresses the quantity of caloric indirectly, as a weight of melted ice; this is what he calls "relative" measure. He will illustrate this practice in a table at (9.10) below.

9.2 *As the substance ... is a solid acid, it is likely that very little caloric remains*: Recall Lavoisier's discussion in Chapter One, where he established that the solid state of bodies is that which contains the least quantity of caloric.

gives us a way of finding out ... the amount of caloric contained in oxygen gas: The oxygen which entered into combustion had been gaseous and is now solid, while the phosphorus was solid to begin with. Therefore most of the caloric released during combustion—the caloric measured in the experiment—probably came from the oxygen gas. Lavoisier will offer similar reasoning in (9.4) and (9.6) below.

that the phosphoric acid still retains a considerable quantity of caloric, since the phosphorus itself contained a certain amount before combustion, the error could only be with respect to the difference, and consequently of little importance.

9.3 I showed in Chapter Five that a livre of phosphorus would absorb 1 livre, 8 ounces of oxygen when burned. And since at the same time 100 livres of ice are melted, it follows from this that the quantity of heat contained in a livre of oxygen gas is able to melt 66 livres, 10 ounces, 5 gros, 24 grains of ice.

9.4 One livre of charcoal when burned only melts 96 livres, 8 ounces of ice. But it absorbs at the same time 2 livres, 9 ounces, 1 gros, 10 grains of oxygen [105] gas. Now reckoning from the results obtained in the combustion of phosphorus, 2 livres, 9 ounces, 1 gros, 10 grains of oxygen gas should give off enough heat to melt 171 livres, 6 ounces, 5 gros of ice. Thus, in this experiment a quantity of caloric which would have sufficed to melt 74 livres, 14 ounces, 5 gros of ice disappears. But as carbonic acid, unlike phosphoric acid, is not at all in a solid state after combustion, but, on the contrary, in a gaseous state, a quantity of caloric was required to bring it to this state, and it is this quantity which is found missing in the above-mentioned combustion. Dividing it by the number of livres of carbonic acid formed by the combustion of a livre of charcoal, we find that the quantity of caloric necessary to bring a livre of carbonic acid from the solid to the gaseous state would melt 20 livres, 15 ounces, 5 gros of ice.

9.3 *a livre of phosphorus would absorb 1 livre, 8 ounces of oxygen when burned*: See (5.13), where Lavoisier gives the ratio of phosphorus to oxygen as roughly 2:3.

9.4 *the quantity of caloric necessary to bring a livre of carbonic acid from the solid to the gaseous state*: Let us restate Lavoisier's calculation in decimal form (see the Appendix for this procedure). When burned, 1 livre of phosphorus absorbs 1.5 livres of oxygen while melting 100 livres of ice (9.3). If the heat necessary for melting came predominantly from the oxygen (9.2 and comment), then 1.5 livres oxygen contain enough heat to melt 100 livres of ice, so that 1 livre of oxygen contains heat sufficient to melt $100 \div 1.5 = 66.667$ livres of ice.

Now 1 livre of charcoal when burned melts 96.5 livres of ice and absorbs 2.571 livres of oxygen. But this much oxygen represents enough heat to melt 2.571×66.667 or 171.401 livres of ice—a discrepancy of $171.401 - 96.5$ or 74.901 livres of ice. This quantity of caloric will have been absorbed by the carbonic acid in assuming the gaseous state. And since burning 1 livre of charcoal forms 3.571 livres of carbonic acid gas (5.24), then caloric sufficient to melt 74.901 livres

9.5 A similar calculation can be made for the combustion of hydrogen and the formation of water. A livre of this elastic fluid absorbs 5 livres, 10 ounces, 5 gros, 24 grains of oxygen when it burns, and melts 295 livres, 9 ounces, $3\frac{1}{2}$ gros of ice.

9.6 Then, according to the results obtained by the combustion of phosphorus, 5 livres, 10 ounces, 5 gros, 24 grains of oxygen gas, passing from an aeriform to a solid state, would lose enough caloric to melt a quantity of ice equal to [106]

	livres	ounces	gros
	377	12	3

9.7 The quantity of ice melted by this oxygen gas when burned with hydrogen gas is only 295 2 3

9.8 Thus, even though the water formed is restored to a temperature of zero, there remains in this water enough caloric to melt 82 9 $7\frac{1}{2}$

9.9 Now, since 6 livres, 10 ounces, 5 gros, 24 grains of water are formed in the combustion of a livre of hydrogen gas, it follows that in each livre of water, at zero degrees, there remains a quantity of caloric equal to that required to melt 12 livres, 5 ounces, 2 gros, 48 grains of ice, without even mentioning that contained in the hydrogen gas. And this is impossible to take into account in this experiment, as we do not know its quantity. From this we can see that water, even in the state of ice, still contains a great deal of caloric and that oxygen conserves a considerable amount of caloric when it combines to form water.

---

of ice must have entered into 3.571 livres of carbonic acid gas. Thus to bring a single livre of carbonic acid to the gaseous state requires caloric sufficient to melt 74.901 ÷ 3.571 = 20.975 livres of ice.

9.5 *this elastic fluid*: that is, hydrogen.

9.9 *oxygen conserves a considerable amount of caloric*: It may be useful here to recall the distinction Lavoisier made in Chapter One between "free caloric" (1.37) and "combined caloric" (1.38). Free caloric has no bond; it passes freely into and out of bodies, which in turn exhibit increasing or decreasing temperature. Combined caloric, by contrast, has a more fundamental connection with other substances; indeed, Lavoisier frequently speaks of the "affinity" particular substances have for caloric—the term which also names the tendency of ordinary substances to combine with one another. In (5.8), for example, he gauges the affinity of oxygen for caloric relative to its affinity for phosphorus. In the case of gases, there is a definite quantity of caloric associated with the gaseous state (9.3, 9.4); but as he shows in the present paragraph, even solids "conserve" a characteristic quantity of this remarkable substance.

9.10 The results of these various trials are summarized as follows:

Combustion of Phosphorus

	livres	ounces	gros	grains
Quantity of phosphorus burned	1			
Quantity of oxygen gas necessary for combustion	1	8		
Quantity of phosphoric acid obtained	2	8		

Quantity of caloric released by combustion of a livre of phosphorus, expressed in terms of the number of livres of ice that can be melted during the combustion	100.00000
Quantity of caloric released from each livre of oxygen gas in the combustion of phosphorus	66.66667
Quantity of caloric released when a livre of phosphoric acid forms	40.00000
Quantity of caloric remaining in each livre of phosphoric acid	0.00000

9.11 Here it is assumed that the phosphoric acid does not retain any portion of its caloric, though this is not strictly true. But (as already observed above) the quantity of caloric is probably very little; and lacking the means to calculate it, we have assumed there is none.

Combustion of Charcoal

	livres	ounces	gros	grains
Quantity of charcoal burned	1			
Quantity of oxygen gas absorbed during combustion	2	9	1	10
Quantity of carbonic acid formed	3	9	1	10

Quantity of caloric released by combustion of a livre of charcoal, expressed in terms of the number of livres of ice that can be melted during the combustion	96.50000
Quantity of caloric released from each livre of oxygen gas	37.52823
Quantity of caloric released when a livre of carbonic acid forms	27.02024
Quantity of caloric remaining in each livre of oxygen after this combustion	29.13844
Quantity of caloric necessary to convert a livre of carbonic acid into a gas	20.97960

9.10 (Table) *Quantity of caloric ... expressed in terms of the number of livres of ice that can be melted*: An example of "relative" measure of caloric (9.1).

9.11 *as already observed above*: In (9.2), for example.

Combustion of Hydrogen Gas [109]

	livres	ounces	gros	grains
Quantity of hydrogen burned	1			
Quantity of oxygen gas employed in the combustion	5	10	5	24
Quantity of water formed	6	10	5	24

Quantity of caloric released by combustion of a livre of hydrogen gas	295.58950
Quantity of caloric released from each livre of oxygen gas	52.16280
Quantity of caloric released when a livre of water forms	44.33840
Quantity of caloric which each livre of oxygen retains after this combustion of hydrogen	14.50386
Quantity of caloric which a livre of water retains at 0 degrees	12.32823

On the Formation of Nitric Acid

9.12 When nitrous gas is combined with oxygen gas to form nitric or nitrous acid, a small amount of heat is produced. But this amount is much less than that which is produced in the other combinations with oxygen. From this it [110] follows as a necessary consequence that oxygen gas, as it becomes a fixed part of nitric acid, retains a large portion of the caloric which was combined with it in the gaseous state. It is by no means impossible to ascertain the quantity of caloric released when these two gases join together, and from this the amount that compound retains could be easily inferred. The first of these givens might be obtained by combining nitrous gas with oxygen gas in an apparatus surrounded by ice. But as very little caloric is released when they combine, the quantity could be successfully determined only by proceeding on a large scale, using cumbersome and overly complicated equipment. And this is precisely what has prevented Laplace and me from attempting it in the past. In the meantime, we can already remedy this with certain calculations which cannot greatly deviate from the truth.

9.13 Laplace and I caused a violent explosion in an apparatus made of ice containing a suitable mixture of saltpeter and charcoal. We observed that in this explosion, a livre of saltpeter was able to melt 12 livres of ice.

9.14 But a livre of saltpeter, as will be seen in what follows, contains the [111] following quantities of potash and dry acid:

9.14 *dry acid*: Here and in the table on the next page, this phrase refers to charcoal, that is, the solid base of carbonic acid.

	ounces	gros	grains	total weight in grains
Potash	7	6	51.84	4515.84
Dry Acid	8	1	20.16	4700.16

9.15 And the 8 ounces, 1 gros, 20.16 grains of acid are in turn composed of:

	ounces	gros	grains	total weight in grains
Oxygen	6	3	66.34	3738.34
Noxious Gas	1	5	25.82	961.82

9.16 In this process there has actually been burned 2 gros, $1\frac{1}{3}$ grain of charcoal, along with 3738.34 grains or 6 ounces, 3 gros, 66.34 grains of oxygen. And since the quantity of ice melted during this combustion was 12 livres, it follows that: one livre of oxygen gas, burned in the same way, would melt   29.58320

9.17 Adding to this the quantity of caloric conserved by a livre of oxygen in its combination with charcoal in order to form carbonic acid in the gaseous state, which is, as was seen above,   29.13844

9.18 We have for the total quantity of caloric contained by a livre of oxygen when combined in nitric acid,   58.72164

[112]   9.19 It is apparent from the result of the combustion of phosphorus that oxygen, in a gaseous state, contains at least   66.66667

9.20 Therefore, when combining with azote to form nitric acid, it loses only   7.94502

9.21 Further experiments will inform us whether this result, which was deduced by calculation, agrees with more direct procedures.

9.22 This enormous quantity of caloric, which oxygen gas carries with it in nitric acid, explains why there is such a great release of caloric in all explosions involving nitre, or to speak more precisely, in all circumstances in which nitric acid decomposes.

Combustion of the Candle

9.23 After having examined several cases of simple combustion, I will give examples of some that are more complex, beginning with wax.

9.20 *it loses only 7.94502*: That is, the *oxygen*, combining with azote to form nitric acid, lost a quantity of caloric sufficient to melt 7.94502 livres of ice.

9.22 *this enormous quantity of caloric*: That is, the quantity of caloric joined to the oxygen in nitric acid: an amount sufficient to melt at least 66.66667 livres of ice.

9.24 A livre of this substance, burning gently in the ice apparatus designed to measure quantities of caloric, melts 133 livres, 2 ounces, $5\frac{1}{3}$ gros of ice.

9.25 But one livre of candle wax, following the experiments I have already published,* contains: [113]

	ounces	gros	grains
Charcoal	13	1	23
Hydrogen	2	6	49

9.26 The 13 ounces 1 gros 23 grains of charcoal, according to the experiments reported above, should melt

livres of ice

79.39390

while the 2 ounces, 6 gros, 49 grains of hydrogen should melt 52.37605

for a total of 131.76995

9.27 These results clearly show that the quantity of caloric released from the burning candle is almost exactly equal to the caloric which would be obtained by separately burning a weight of charcoal and hydrogen equal to the weight that enters into their combination. Since the experiments on the burning candle have been repeated several times, I have grounds for presuming that they are exact.

Combustion of Olive Oil

9.28 We enclosed a lamp containing a determinate quantity of olive oil in the standard apparatus. When the experiment was finished, we established the exact weight of oil that had been consumed, and also the weight of the ice that had been melted. The result was that a livre of burning olive oil could [114] melt 148 livres, 14 ounces, 1 gros of ice.

9.29 But according to the published experiments I just cited (and which are excerpted in the next chapter), one livre of olive oil contains:

	ounces	gros	grains
Charcoal	12	5	5
Hydrogen	3	2	67

The combustion of 12 ounces, 5 gros, 5 grains of charcoal, should melt only

livres of ice

76.18723

and that of 3 ounces, 2 gros, 67 grains of hydrogen, 62.15053

for a total of 138.33776

9.30 The amount of ice melted is 148.88330

9.31 The release of caloric has therefore been considerably greater than it should have been, by a quantity equivalent to 10.54554

---

* *Mémoires de l'Académie*, 1784, p. 606.

9.32 This difference, which is not, after all, very considerable, can be attributed to the inevitable errors in experiments like these, or to the fact that the composition of oil is not yet known with sufficient exactitude. But it still follows that there is already a great deal of conformity and agreement in the series of experiments concerning the combination and release of caloric.

9.33 What still must be done, and what we are now trying to do, is to determine how much caloric oxygen retains when it combines with metals and converts them to oxides, and how much caloric hydrogen contains in the different states in which it occurs, and finally to learn how much caloric is released during the formation of water. There remains such a degree of uncertainty concerning this amount that new experiments are required to determine it. Once these issues have been fully resolved, as we hope they will be soon, we will probably have to correct many of the results I have presented, perhaps quite substantially. But I did not believe that I should consider this a reason to postpone helping those who will be able to step up to work on these issues. When one is seeking the elements of a new science, it is difficult to begin with anything but rough approximations, and it is rarely possible with a first sketch to bring a new science to a state of perfection.

9.32 *not, after all, very considerable*: The excess (about 10.5 livres) represents 7.5% of the expected quantity (about 138.3 livres). But this is the very quantity Lavoisier called "considerably greater than it should have been" in the previous paragraph. Has he now altered his assessment? If so, why?

Chapter Ten

On the Combination of Combustible Substances with One Another

10.1 Since combustible substances in general have a great appetite for oxygen, it follows that they must also have affinities for one another and that they must tend to combine with each other: *Things which are identical to some other thing are identical to one another.** And in fact this is what we observe. Nearly all metals, for example, are able to combine with one another, resulting in an array of composites customarily referred to as "alloys." Nothing prevents us from adopting this expression. Thus we will state that most metals "form alloys" with one another; that like all compounds, alloys admit of one or more degrees of saturation; and that metallic substances in this state are in general more brittle than pure metals, especially when the metals that form the alloy have very different melting points. Finally, we will add that these different metallic melting points account for some of the phenomena peculiar [117] to alloys, for example, the property that several varieties of iron have of being brittle when heated. These irons must be considered as alloys of pure iron, an almost infusible metal, with a small quantity of some other metal, whatever it may be, which becomes liquid at a much lower heat. So long as an alloy of this kind is cold and the two metals are in a solid state, it is malleable. But if heated to a degree sufficient to melt the more fusible of the two metals, the liquid parts interposed between the solid ones are bound to interrupt the continuity of the mixture and force the iron to become brittle.

* [This sentence appears in Latin in the French edition: *quae sunt eadem uni tertio sunt eadem inter se.* —Tr.]

0.1 *Since combustible substances ... have a great appetite for oxygen, it follows that they must also have affinities for one another*: Why does this "follow"? Lavoisier appears to deduce it from the general principle that "things which are identical to some other thing are identical to one another." Note that if sharing such an "appetite" really does constitute being identical, then chemical affinity will contrast sharply with phenomena like electrical or magnetic action, in which opposites attract while likes repel.

10.2 Alloys of metals with mercury are customarily called "amalgams," and we have seen no inconvenience in preserving this denomination.

10.3 Sulphur, phosphorus, and charcoal are also capable of combining with metals. Compounds involving sulphur have in general been named "pyrites." The others have not been named, or at least they have received such modern denominations that nothing stands in the way of changing them.

[118]   10.4 We have given the name of "sulphurets" to the first of these compounds, "phosphurets" to the second, and lastly "carburets" to the third. Thus oxygenated sulphur, phosphorus, and charcoal form oxides or acids. But when they combine without being previously oxygenated, they form sulphurets, phosphurets, and carburets. We will even extend these denominations to alkaline compounds. Thus we will give the name "sulphuret of potash" to the compound of sulphur with potash, or fixed vegetable alkali, and the name "sulphuret of ammoniac" to the compound of sulphur with volatile alkali (ammoniac).

10.5 That exceptionally combustible substance, hydrogen, is also capable of combining with a great number of combustible substances. In the gaseous state, hydrogen dissolves carbon, sulphur, phosphorus, and several metals. We will name these combinations "carbureted hydrogen gas," "sulphureted hydrogen gas," and "phosphorated hydrogen gas." The second of these gases, sulphureted hydrogen gas, is the one that chemists named "hepatic gas," and Scheele named "stinking gas from sulphur." Several mineral waters owe their

[119]   powers to this gas. In addition, animal excrements owe their putrid odors primarily to its emissions. With regard to phosphorated hydrogen gas, it has the remarkable property of igniting spontaneously when in contact with air, or better yet, when in contact with oxygen gas, as Gengembre discovered. This gas has the smell of rotten fish, and in fact it is likely that putrifying fish flesh emits a true phosphorated hydrogen gas.

10.6 When hydrogen and carbon unite without the caloric bringing hydrogen into a gaseous state, the result is a particular compound known by the name of "oil"; and this oil is either fixed or volatile, according to the proportion of hydrogen and carbon.

10.6 *When hydrogen and carbon unite ... the result is "oil"*: Lavoisier has not yet stated how he knows that oils consist of hydrogen and carbon exclusively.

this oil is either fixed or volatile...: "Fixed" oils tend to remain liquid at ordinary temperatures, while "volatile" oils readily evaporate.

...according to the proportion of hydrogen and carbon: In the next paragraph Lavoisier will assert that fixed oils embody a greater proportion of carbon than do volatile oils.

10.7 It will not be useless to note here that one of the principal characteristics which distinguish fixed oils pressed out of vegetables from the volatile or essential oils is that the first contain an excess of carbon, which separates off when they are heated above the temperature of boiling water. The volatile oils, on the contrary, are formed with a better balance of carbon to hydrogen, and are not at all capable of being decomposed at a degree of heat greater than that of boiling water. The two principles that constitute them remain united: they combine with caloric to form a gas, and it is in this state that the oils are distilled. [120]

10.8 I have already published a proof that oils are in this way composed of hydrogen and carbon in a memoir on the combination of alcohol and oils with oxygen.* It will be seen there that fixed oils, burning in oxygen gas, convert

* *Recueil de l'Académie*, 1784, p. 593.

10.7 *fixed oils pressed out of vegetables…*: Lavoisier here emphasizes the process rather than the source, for both fixed and volatile oils can be obtained from plant matter. But fixed oils are obtained by pressing—olives, for example—while volatile oils are ordinarily procured through distillation of leaves or flowers.

…contain an excess of carbon…: How does Lavoisier know this? We will find one possible reason in the next paragraph; a second will emerge in Chapter Twelve.

…which separates off: Fixed oils, then, *decompose* at high temperatures, while volatile oils do not. Wouldn't one expect the opposite? But remember that the epithets "fixed" and "volatile" refer not to decomposition but to *vaporization*: fixed oils tend to remain liquid at ordinary temperatures, while volatile oils evaporate quickly (10.6, comment).

volatile oils [exhibit] a better balance of carbon to hydrogen: It is reasonable to expect that a constituent present to an excessive degree would be easily driven off. Volatile oils, then, being more stable under the action of heat than fixed oils, probably enjoy a better balance of components, implying that fixed oils contain an superabundance of one of them; and since it is *carbon* that is "separated off," carbon is probably the excess ingredient. But what does such separation look like? Lavoisier will give some examples in Chapter Twelve, but they are not as specific as we might wish.

a gas … it is in this state that the oils are distilled: Lavoisier is still speaking only of volatile oils. The products of distillation are driven off in gaseous form by heat, and subsequently condensed by cooling to the liquid state. As Lavoisier will outline more fully in Chapter Twelve, volatile oils are formed by the very process of distillation, in contrast to fixed oils, which exist already in their vegetable source. Moreover, volatile oils may be purified by a second or third distillation; but fixed oils are unable to withstand the high temperatures that distillation requires.

into water and carbonic acid. The calculated results of the experiment show that the oils are composed of 21 parts hydrogen and 79 parts carbon. Perhaps solid, oily substances, such as wax, also contain a little oxygen, to which they owe their solid state. I am, moreover, occupied at the moment with experiments which will greatly develop this whole theory.

10.9 It would be well worth determining whether hydrogen, in its pure and uncombined state, is capable of combining with sulphur, phosphorus, and even with metals. Nothing whatsoever indicates *a priori* that these

---

10.8 *The calculated results of the experiment show the oils are composed of 21 parts hydrogen and 79 parts carbon*: These results from combustion of fixed oils would seem to constitute Lavoisier's grounds for asserting that oils consist solely of hydrogen and carbon; but the situation is not quite so straightforward.

As there reported, Lavoisier burned 19.25 grains of olive oil; its combustion consumed 62.00 grains of oxygen gas and produced 27.00 grains of water and 54.25 grains of carbonic acid exclusively. Since he knows that carbonic acid is 72% oxygen (5.23), 54.25 × .72 = 39.06 grains of oxygen were supplied by (and therefore must have been contained in) the oil. Then the remaining oxygen, 62.00 − 39.06 = 22.94 grains, must have gone to the water. But he knows that water is 15% hydrogen (8.16); so 27.00 × .15 = 4.05 grains of hydrogen must have been supplied by (and hence contained in) the oil. Conclusion: olive oil is 21% (4.05 ÷ 19.25) hydrogen and 79% (100−21) carbon. *But this reasoning assumes the very point in question,* namely, that oil contains nothing other than hydrogen and carbon; and in particular—when subtracting 39.06 from 62.00 above—that it contains no oxygen, so that all the oxygen in the products was supplied by the 62 grains that were consumed.

Since the foregoing reasoning depends on a questionable assumption, what is Lavoisier's actual justification for the 21:79 ratio? Frederick Holmes has put forward an ingenious suggestion. In Chapter 10 of his *Lavoisier and the Chemistry of Life, An Exploration of Scientific Creativity* (University of Wisconsin Press, 1987), Holmes points out that in Lavoisier's notebook there appears an alternate hydrogen calculation which, while still assuming hydrogen and carbon as the sole constituents, does *not* assume the 15:85 composition for water. When the quantities of hydrogen and oxygen calculated in this fashion are compared, the 15:85 ratio is confirmed—something that could not happen if the oil had contained oxygen. So we can, after all, accept that this fixed oil embodies 79 parts carbon and 21 parts oxygen exclusively.

Nevertheless, Lavoisier did not undertake a similar analysis of a volatile oil; so we still lack direct evidence that volatile oils contain proportionally less of carbon.

a little oxygen, to which they owe their solid state: Why would Lavoisier think that oxygen lends solidity? Perhaps because he found earlier that some nonmetal oxides (Chapter Five) and all metal oxides (Chapter Seven) are solid. On the other hand, the oxides of sulphur and carbon are not; so it must remain a conjecture.

combinations are impossible. Since combustible bodies are in general capable of combining with one another, it is not clear why hydrogen would be an [121] exception. But at the same time, no direct experiment as yet proves either the possibility or impossibility of this union. Among all the metals, iron and zinc are the ones which we would most reasonably expect to combine with hydrogen. But at the same time, these metals have the property of decomposing water. And as it is difficult in chemical experiments to get rid of the last traces of moisture, it is not easy to ascertain whether the small portions of hydrogen gas obtained in several experiments on these metals were previously combined with them, or whether they come about from the decomposition of a few particles of water. What is certain is that the more care that is taken to remove water in experiments like these, the more the quantity of hydrogen gas diminishes. If very great precautions are taken, only an almost imperceptible quantity of this gas is obtained.

10.10 Whether or not combustible bodies, especially sulphur, phosphorus, and the metals, can absorb hydrogen, it is at least certain that only a very small quantity combines with those bodies and that the combination with hydrogen, far from being essential to the constitution of these substances, [122] can only be considered as a foreign addition which alters their purity. It is, moreover, up to those who have embraced this position to prove the existence of this hydrogen by decisive experiments; but up until now they have only made conjectures supported by suppositions.

10.9 *the ones which we would most reasonably expect to combine with hydrogen*: Because iron and zinc are especially active metals (3.16, 3.17, 7.10 and comments).

these metals have the property of decomposing water: Lavoisier described the decomposition of water by iron in Chapter Eight.

experiments like these: That is, experiments which seem to obtain hydrogen from metals alone. But since almost no hydrogen appears when great care is taken to prevent the presence of water in the experiment, it is likely that the hydrogen so obtained came exclusively from a small residue of water; and any hydrogen associated with metals should therefore be regarded as an impurity rather than an essential constituent of them.

10.9 *those who have embraced this position*: That is, those chemists who believe hydrogen to be an essential constituent of sulphur, phosphorus, and metals. Lavoisier names several such in Part II of the Treatise.

Chapter Eleven

Remarks on Oxides and Acids with Several Bases, and on the Composition of Vegetable Matter and Animal Matter

11.1 In Chapters Five and Eight, we examined the results of the combustion or oxygenation of the four simple, combustible substances: phosphorus, sulphur, carbon, and hydrogen. We showed in Chapter Ten that the simple, combustible substances were capable of combining with one another to form compound, combustible bodies, and we observed that oils in general, chiefly the fixed oils from vegetables, belong to that class and that they are all composed of hydrogen and carbon. In this chapter, it remains for me to discuss the oxygenation of compound, combustible bodies, to show that there exist acids and oxides with double and triple bases, and that nature supplies us with examples of this every day, and that it is mainly by means of combinations like these that she has been able to achieve such a great variety of results with so few elements or simple bodies.

11.2 It had been observed long ago that the result of blending muriatic and nitric acid together was a mixed acid, having very different properties from those of the two acids of which it was composed. The acid was celebrated for the property it has of dissolving gold, the "king of metals" in the language of alchemy, and for this reason it has been given the illustrious title of "aqua regia." This mixed acid, as has been very thoroughly proven by Berthollet, has specific properties which depend upon the combined action of its two acidifiable bases. This is why we believed we ought to reserve a special name for it. The name of nitro-muriatic acid seemed most suitable to us, because it expresses the nature of the two substances that compose it.

11.3 But although nitro-muriatic acid was the only acid observed to have a double base, this phenomenon presents itself constantly throughout the vegetable kingdom. Simple acids, that is to say, acids composed of only a single acidifiable base, are exceptionally rare there. All the acids in this kingdom have hydrogen and carbon for bases, occasionally hydrogen, carbon, and phosphorus, the whole combined with a greater or lesser proportion of

[124]

[125]

11.3 *this phenomenon*: That is, the existence of substances having multiple bases.

oxygen. The vegetable kingdom also has oxides which are formed from the same double and triple bases, but less oxygenated.

11.4 The acids and oxides of the animal kingdom are even more compounded. For the most part, they contain the four acidifiable bases: hydrogen, carbon, phosphorus, and azote.

11.5 I will not expand much further here on this matter, about which I have only very recently formed clear and systematic ideas. I will discuss it in greater depth in the memoirs that I am preparing for the Academy. Though most of my experiments are complete, I must repeat them and multiply them further so as to be able to give exact results for the quantities. Consequently, I will content myself here with a brief enumeration of the vegetable and animal oxides and acids, and several concluding reflections on the constitution of animals and vegetables.

11.6 The vegetable oxides with two bases are sugar, the different kinds of gums which we have collected under the generic name *mucus*, and starch. These three substances have hydrogen and carbon combined together as [126] their radical so as to form a single base, and are brought to the oxide state by a measure of oxygen. They only differ with respect to the proportions of the principles which compose their base. They can be made to pass from the oxide state to that of an acid by combining with a fresh quantity of oxygen. Thus the different vegetable acids are formed according to the degree of oxygenation and the proportion of hydrogen to carbon.

11.7 To apply the principles previously established for mineral oxides and acids to the naming of vegetable acids and minerals would be nothing more than to assign them names referring to the two substances which compose their bases. The vegetable oxides and acids would then be hydro-carbonous oxides and acids. Additionally, this method would have the advantage of allowing us to indicate without circumlocution which principle is in excess, like the method Rouelle invented for vegetable extracts. He named something "extracto-resinous" when the extract was in excess and "resino-extractive" when there was a greater share of resin.

11.8 Setting forth from the same principles and varying the endings so as to give this language even greater scope, suggests the following names for the [127] vegetable acids and oxides:

Hydro-carbonous oxide
Hydro-carbonic oxide
Carbono-hydrous oxide
Carbono-hydric oxide

Hydro-carbonous acid
Hydro-carbonic acid
Oxygenated hydro-carbonic acid
Carbono-hydrous acid
Carbono-hydric acid
Oxygenated carbono-hydric acid

11.9 It is likely that this language will provide us with a variety of expressions sufficient for the variety with which nature presents us, and that as the vegetable acids become better known they will arrange themselves naturally and, so to speak, by themselves, within the framework we have just supplied. But we are still far from being able to classify these substances in a systematic way. We know the principles which compose them; I no longer have any doubt in this regard. But the proportions of the principles are still unknown. These considerations have caused us to preserve temporarily the former [128] names. And even though I am now further advanced in my researches than I was at the time when our essay on nomenclature first appeared, I would reproach myself if I too confidently inferred conclusions from experiments that are not yet sufficiently exact. But while I admit that this branch of chemistry remains at a standstill, I can add my hope that it will soon be illuminated.

11.10 I feel myself even more decisively compelled to take the same position with respect to the oxides and acids having three and four bases, of which the animal kingdom offers a great number of examples, and which are sometimes encountered even in the vegetable kingdom. Azote, for example, enters into the composition of prussic acid; it is found there joined with hydrogen and carbon, forming a triple base. It likewise enters into gallic acid, as far as we can tell. Finally, nearly all animal acids have azote, phosphorus, hydrogen, and carbon for a base. A system of naming which undertook the task of expressing all four of these bases at the same time would be methodical, no doubt: it would have the advantage of expressing clear and determinate ideas. But this heaping together of Greek and Latin substantives and adjectives, which even chemists [129] do not generally use, would seem to produce a barbarous language, as difficult to retain as to pronounce. Besides, the perfection of the science must precede the perfection of language, and this part of chemistry is far from having

---

11.10 *the perfection of the science must precede the perfection of language:* But recall that in his Preliminary Discourse, Lavoisier argued that the dependence holds in both directions, for "one cannot perfect the language without perfecting the science, nor the science without the language" (0.3).

reached the point to which it must one day arrive. It is therefore indispensable to preserve, at least for a while, the older names for the animal acids and oxides. We have permitted ourselves only to make a few minor modifications: for example, in the case of those substances where we suspect the acidifiable principle is in excess, to use the ending "ous," and, on the contrary, in the case of those where we have reason to believe that oxygen is predominant, to use the ending "ic."

11.11 The vegetable acids currently known are thirteen in number, namely:

Acetous acid
Acetic acid
Oxalic acid
Tartarous acid
Pyro-tartarous acid
Citric acid
Malic acid
Pyro-mucous acid
Pyro-lignous acid
Gallic acid
Benzoic acid
Camphoric acid [130]
Succinic acid

11.12 Although all of these acids, as I have said, are chiefly and almost exclusively composed of hydrogen, carbon, and oxygen, they do not, speaking properly, contain either water, carbonic acid, or oil, but rather only the principles suitable for forming them. The attractive force that hydrogen, carbon, and hydrogen exert amongst themselves is in a state of equilibrium that can exist only at the temperature in which we live. For if they are heated to a degree only a little higher than that of boiling water, the equilibrium is disturbed; the oxygen and water reunite to form water; a portion of the carbon unites with hydrogen to produce oil, and carbonic acid is formed by the combination of carbon and oxygen; and, finally, a surplus quantity of charcoal nearly always remains free. This is what I propose to develop somewhat further in the following chapter.

11.13 Still less is known about the oxides from the animal kingdom than about those from the vegetable kingdom; even their number is yet to be determined. The red part of blood, lymph, and nearly all secretions are true oxides, and it is important to study them from this point of view. [131]

11.14 As for the animal acids, the number of those that are known is actually limited to six. It is still probable that several of these are of the same kind, or at least only differ from one another in a manner that is hardly discernable. These acids are:

> Lactic acid
> Saccho-lactic acid
> Bombic acid
> Formic acid
> Sebacic acid
> Prussic acid

11.15 I do not list phosphoric acid in the rank of animal acids, because it belongs equally to all three kingdoms.

11.16 The connection between the principles that constitute the animal acids and oxides is not stronger than that between the vegetable acids and oxides. A very small change in temperature suffices to disturb it. I hope to make this more apparent through observations I am going to report in the following chapter.

11.15 *all three kingdoms*: Namely, animal, vegetable, and mineral.

Chapter Twelve

On the Decomposition of Vegetable and Animal Matter by the
Action of Fire

12.1 To have a proper conception of what happens during the decomposition of vegetable substances by fire, one must not only consider the nature of the principles that compose these substances, but also the different attractive forces that the particles of these principles exert on one another and. at the same time, the attractive force that caloric exerts on the particles.

12.2 As I just explained in the preceding chapter, the constitutive principles of vegetables are reduced to three: hydrogen, oxygen, and carbon. I call them "constitutive" because they are common to all vegetables, and vegetables could not exist without them. This is in contrast with other substances that are essential only to the constitution of a particular vegetable and not of all vegetables in general.

12.3 Two of these three principles, hydrogen and oxygen, have a great tendency to unite with caloric and convert to gases. Carbon, on the contrary , is a fixed principle and has very little affinity for caloric.

12.4 On the other hand, oxygen, which tends to unite with either hydrogen or carbon with a nearly equal force at the usual temperature in which we live, has more affinity for carbon when red hot. As a result, at this degree of heat oxygen leaves hydrogen and unites with carbon, forming carbonic acid.

12.5 I will sometimes make use of the expression "red heat." While it does not express a determinate degree of heat, it nevertheless does express a degree of heat much greater than that of boiling water.

12.6 Although we are far from knowing the strength of all these forces and from being able to express their intensity with numbers, we are at least certain, by what occurs on a daily basis before our eyes, that however variable they may be on account of the degree of temperature or, what is the same thing, on account of the quantity of caloric with which they are combined, they are all nearly in equilibrium at the temperature in which we live. Thus

[134] vegetables contain neither oil, water, nor carbonic acid,* but they do contain the elements of all of these substances. Hydrogen does not combine at all with oxygen alone nor with carbon alone, and inversely. But the particles of these three substances form a triple compound, and are as a result at rest and in equilibrium.

12.7 A very slight change in temperature is enough to topple this "scaffolding of combinations," if I may use this expression. If the temperature to which the vegetable matter is exposed is not much greater than that of boiling water, hydrogen and oxygen reunite and form water which undergoes distillation. A portion of the hydrogen and carbon unite to form volatile oil; another portion of carbon becomes free, and, being the most fixed of the principles, it remains in the retort. But if, instead of a heat near that of boiling water, we apply red heat to a vegetable substance, water is now no longer formed; or rather, even the water that was able to form during the initial contact with the heat decomposes. Oxygen unites with carbon, with which it has more affinity at this degree of temperature. Carbonic acid forms; and the hydrogen, now

---

* It is clear that I here assume that the vegetables are reduced to a state of perfect dryness and that, with regard to oil, I am not speaking about vegetables which furnish it either by a cold press or by a heat which does not exceed that of boiling water. Here it is only a question of the empyreumatic oil obtained through distillation over an open fire at a degree of heat greater than that of boiling water. It is only this oil that I declare to be the result of the process. What I published in this regard may be seen in the volume of the Academy for 1786. [This footnote was added in the 1793 edition. —Tr.]

12.6 *neither oil, water, nor carbonic acid, but ... the elements of all these*: This is misleading. In the footnote added to this paragraph for the 1793 edition, Lavoisier acknowledges that the oils obtained by pressing or by low-temperature distillation *do* exist ready-formed in the plant material, since no decomposition takes place in those processes. In contrast, oils obtained through high-temperature distillation result from recombination of constituents that are liberated when vegetable matter decomposes at elevated temperatures; therefore these oils do not exist prior to the distillation process; only their constituents do. Lavoisier will say more about the sequence of liberation and recombination in the following paragraphs.

12.7 *another portion of carbon becomes free*: Carbon "separates off," which Lavoisier took as evidence of its "excess" in (10.7). Notice, therefore, that the quantity of carbon which enters into the volatile oil produced in this process is *less* than the vegetable matter contained. That is perhaps a more direct indication that volatile oils contain proportionally less carbon then do fixed oils.

... being the most fixed of the principles ... remains in the retort: That is, the carbon does not vaporize; the epithet "fixed" evidently has the same meaning here as it did in the case of "fixed oils" (10.6).

free, unites with caloric and escapes in the form of a gas. At this temperature, not only is no oil formed, but also any that had been previously formed would be decomposed. [135]

12.8 We see, therefore, that vegetable matter is decomposed at this temperature by the play of double and triple affinities and that, while carbon attracts oxygen to form carbonic acid, caloric attracts hydrogen to form hydrogen gas.

12.9 There is no vegetable substance whatever whose distillation does not provide proof of this theory—if, that is, we can call a simple statement of facts a "theory." Let sugar be distilled. As long as it is only subjected to a heat less than that of boiling water, it will lose only a small amount of its water of crystallization. It will remain sugar, preserving all of its properties. But as soon as it is exposed to a heat even slightly greater than that of boiling water, it turns black. A portion of the carbon separates from the combination, and slightly acidified water, with a little oil, is distilled. The carbon that remains in the retort comprises nearly a third of the original weight.

12.10 The play of affinities is even more complex in plants containing azote, such as crucifers, and in those containing phosphorus. But as these substances are involved in only small quantities in these combinations, they [136] do not bring about great changes, at least in appearance, to the phenomena of distillation. The phosphorus appears to remain combined with the carbon, which imparts fixity to it. As for the azote, it unites with hydrogen to form ammoniac (volatile alkali).

At this temperature ... is no oil formed: Both here and in the next paragraph, "this temperature" denotes the *red heat* which Lavoisier defined in (12.5); and he clearly implies that such intense heat decomposes even the volatile oils that might be obtained though a more moderate distillation. But this contradicts the 1793 footnote, where he represents "empyreumatic" oils (oils having an odor of burnt organic matter) as being produced at high temperatures in just this way.

12.9 *it turns black. A portion of the carbon separates*: The sugar turns black because carbon has been driven from it. The ease with which this happens—requiring a temperature only slightly above that of boiling water—may indicate what Lavoisier called "excess carbon" (10.7). Note that he regards sugar as a vegetable substance; but he will shortly report a similar phenomenon in animal matter (12.11).

12.10 *The phosphorus appears to remain combined with the carbon, which imparts fixity to it*: During distillation, the phosphorus remains part of the plant structure and does not appear among the distillation products. But if that is what constitutes its "fixity"—the quality that particularly distinguishes carbon (12.7)—by what means does Lavoisier suppose carbon to "impart" that quality?

12.11 Since animal substances are composed of nearly the same principles as cruciferous plants, their distillation gives the same results. But as they contain more hydrogen and more azote, they furnish more oil and ammoniac. In order to show how readily this theory accounts for all the phenomena associated with the distillation of animal matter, I will only cite a single fact: the purification and total decomposition of volatile animal oils, commonly called Dippel's oils. When these are obtained by a first distillation over an open fire, these oils are brown, because they contain a small amount of nearly free carbon. But rectification renders them colorless. The carbon adheres so weakly in these compounds that it separates off when they are exposed to air. If volatile animal oil which is well-rectified, and consequently colorless, limpid, and transparent, is placed under a bell jar filled with oxygen gas, the volume of the gas diminishes in a short time, being absorbed by the oil. The oxygen combines with the hydrogen from the oil to form water, which falls to the bottom. At the same time, the portion of carbon which was combined with hydrogen becomes free and manifests itself by its black color. This is the reason why oils can be maintained colorless and clear only when they are contained in tightly closed flasks, and why they blacken as soon as they come into contact with air.

[137]

12.11 *when these are obtained by a first distillation...*: Volatile oils can also be produced by distillation of animal matter, which in the course of that process undergoes decompositions resembling those described for vegetable materials in (12.7).

...these oils are brown, because they contain ... nearly free carbon: Recall from (12.7) that when carbon is liberated from plant matter in distillation, one portion recombines with hydrogen to form volatile oil, while another portion either remains in the distilling vessel or, uniting with oxygen, passes over as carbonic acid. Then how does "nearly free carbon"—itself a puzzling locution—end up among the distillation products?

rectification: That is, a second or even third distillation. Each distillation cycle expels a portion of free carbon as carbonic acid; eventually a colorless oil is obtained.

The carbon adheres so weakly in these compounds that it separates off: It seems, then, that the enigmatic phrase "nearly free carbon" refers to carbon that is weakly bound and therefore separates easily from its compounds. But "these compounds" are *volatile* oils, albeit from animal rather than vegetable sources. This is rather awkward for Lavoisier's theory. A principal reason for his having supposed volatile oils to have a "better balance" between carbon and hydrogen (10.7) was the ease with which fixed oils, but not volatile oils, can be induced to lose carbon. Evidently, though, that contrast between fixed and volatile oils is not universal.

12.12 Successive rectifications of these same oils exhibit another phenomenon that confirms this theory. Each time they are distilled, a small amount of carbon remains at the bottom of the retort. At the same time a little water forms through the combination of oxygen from the air in the vessels with hydrogen from the oil. As the same phenomenon occurs with each distillation of the same oil, it follows that at the end of a great number of successive rectifications, especially if a fairly hot fire and fairly large vessels are used in the procedure, all of the oil is decomposed, successfully converted in its entirety into water and charcoal. This total decomposition of oil by repeated rectification takes much longer and is much more difficult when vessels with small capacity are used, and especially if the degree of heat is low and hardly greater than that of boiling water. I will give a detailed account of my experiments on the decomposition of oils in a separate memoir to the Academy. But what I have said here appears sufficient to convey precise ideas about the constitution of vegetable and animal materials, and on their decomposition by fire.

[138]

Chapter Thirteen

On the Decomposition of Vegetable Oxides through Vinous Fermentation

13.1 Everyone knows how wine, cider, mead, and, in general, all fermented spirits are made. The juice from grapes or apples is pressed out, diluted with water, and then the liquid is placed in large vats and kept in a place where the temperature is at least 10 degrees, as measured by Réaumur's thermometer. Soon, a rapid motion of fermentation is provoked, numerous air bubbles rise up and burst on the surface; and, when the fermentation is at its peak, the quantity of bubbles released is so great, the quantity of gas released is so enormous, that it might seem as if some blazing fire had excited the liquid to a violent boil. The gas released is carbonic acid, and when collected with care, it is perfectly pure and free from mixture with all other kinds of air or gas.

13.2 In this process, the grape juice, which was initially sweet and sugary, becomes a vinous liquor. Once the fermentation is complete, it no longer contains any sugar, and an inflammable liquor known to merchants, scientists and artisans as "spirit of wine" can be distilled from it. But since this liquor can be produced by fermenting any sugary matter that has been sufficiently diluted with water, it is clear that it would have been contrary to the principles of our nomenclature to call it "spirit of wine" rather than "spirit of cider" or "spirit of fermented sugar." So we were compelled to adopt a more general term; and "alcohol," a term derived from Arabic, seemed appropriate for our purposes.

[140]

13.3 Fermentation is one of the most striking and extraordinary processes in chemistry. We need to determine the source of both the carbonic acid gas which is released and the inflammable spirit which is formed, and examine how a body that is sweet, a vegetable oxide, can be transformed in this way into two such different substances, one combustible and the other exceptionally incombustible. To arrive at a solution to these two questions, it is clearly necessary, first of all, to understand fully both the analysis and the nature of the body capable of fermentation, and also the products of fermentation. For nothing is created, either in the processes of art or in those of

[141]

nature. And one can establish as a rule that in every process there is an equal quantity of matter both before and afterwards; that the quality and quantity of the principles is the same: there are only changes or modifications.

13.4 These are the rules on which the entire art of experimentation in chemistry is founded. In chemistry, perfect equality, or an equation, between the principles of the body examined and those principles that are obtained from them by analysis must always be assumed. Thus, since the must of grapes produces carbonic acid gas and alcohol, I can say that "must of grapes = carbonic acid + alcohol." It follows from this that there are two approaches to explaining what happens during vinous fermentation. The first is to determine accurately the nature and the principles of the fermentable bodies. The second is to observe carefully the results that fermentation produces. It is evident that whatever we learn about the objects of analysis will have consequences for the nature of the products of analysis, and *vice versa*.

13.5 Because of this, it was important for me to strive for a good understanding of the constitutive principles of fermentable bodies. It is clear that in order to achieve my goal I did not have to seek out highly complex fruit juices [142] which would have been impossible, perhaps, to analyze rigorously. Instead, I chose the most simple of those bodies capable of fermentation, that is, sugar, which is easy to analyze and whose nature I have already explained. Recall that this substance is a true vegetable oxide, an oxide with two bases; that it is composed of hydrogen and carbon brought to the state of an oxide by a

---

13.3 *in every process there is an equal quantity of matter both before and afterwards*: This is the fundamental principle of Lavoisier's experimental art, as indeed he will state in the next paragraph. It is the reason why precise measurement of the *weights* of reactants and products is so important. The principle has been central to Lavoisier's practical regimen since Chapter Three (3.25, comment); why has he waited until now to formulate it explicitly?

the quality and quantity of the principles is the same: Notice that, while the previous statement asserted constancy only of the *total quantity of matter*, the present dictum extends to each "principle" separately.

there are only changes or modifications: If the *principles* do not change either in quality or quantity, what is it that undergoes "changes or modifications"?

13.4 *must of grapes = carbonic acid + alcohol*: What significance should attach to Lavoisier's sudden recourse to algebraic formalism here? Notice that the equation form carries a double rhetoric. First, it expresses *quantitative equality* (of weights), as indeed the very term "equation" requires. But just as important, it expresses *qualitative identity*, declaring that must of grapes *is* a combination of carbonic acid and alcohol.

certain proportion of oxygen; and that these three principles are in a state of equilibrium that a very weak force suffices to break. A long series of experiments made in different ways and repeated many times has taught me that the proportions of the principles entering into the composition of sugar are close to the following:

Element	Parts
Hydrogen	8
Oxygen	64
Carbon	28
Total	100

13.6 In order to ferment the sugar, it must first be diluted with around four parts of water. But no matter the proportion, the water and sugar, mixed together, will never ferment by themselves; and the equilibrium between the principles of this compound will remain forever unless that equilibrium is somehow disturbed. A little yeast from beer suffices to produce this effect and to start off the fermentation, which then continues on its own to the end. Elsewhere I will give an account of the effects of yeast and of fermenting agents in general. I usually used ten livres of yeast paste for a hundred livres of sugar, and a quantity of water equal to four times the weight of the sugar. And so the fermentable liquid was composed just as follows. I am giving here the results of my experiments precisely as I obtained them, preserving even the fractions given me by the calculations of the reduction.

[143]

Materials of Fermentation for 100 Livres of Sugar

		livres	ounces	gros	grains
	Water	400			
	Sugar	100			
Paste of beer yeast,	Water	7	3	6	44
composed of	Dry yeast	2	12	1	28
	Total	510			

13.6 *this effect*: That is, disturbance of equilibrium.

start off the fermentation, which then continues on its own: Note that Lavoisier credits yeast only with the *initiation* of fermentation, not as entering into the process of fermentation itself.

Detailed Account of the Principles Constituting the Materials of Fermentation

[144]

livres	ounces	gros	grains			livres	ounces	gros	grains
407	3	6	44	of water composed of	hydrogen	61	1	2	71.40
					oxygen	346	2	3	44.60
100				of sugar composed of	hydrogen	8			
					oxygen	64			
					carbon	28			
2	12	1	28	of dry yeast composed of	carbon		12	4	59.00
					azote			5	2.94
					hydrogen		4	5	9.30
					oxygen	1	10	2	28.76
			Total			510			

Recapitulation of the Principles Constituting the Materials of Fermentation

		livres	ounces	gros	grains	livres	ounces	gros	grains
Oxygen	from water	340							
	from yeast water	6	2	3	44.60	411	12	6	1.36
	from sugar	64							
	from yeast	1	10	2	28.76				
Hydrogen	from water	60							
	from yeast water	1	1	2	71.40	69	6		8.70
	from sugar	8							
	from yeast		4	5	9.30				
Carbon	from sugar	28				28	12	4	59.00
	from yeast		12	4	59.00				
Azote	from yeast							5	2.94
	Total					510			

13.6 (Table) *Detailed Account of the Principles*: Lavoisier reported the composition of water in (8.28); he determined the makeup of sugar by dry distillation, the process described in (12.9). He does not reveal how he arrived at the composition of dry yeast, but similar distillation measures would have been suitable.

[145]   13.7 After the precise determination of the nature and the quantity of the principles of the material fermented, there remains the examination of the products. To arrive at knowledge of them, I began by enclosing the 510 livres of the liquid mentioned above in an apparatus by means of which I could not only determine the quality and quantity of gases as they were released, but could also weigh each of the products separately, at whatever stage of the fermentation I deemed appropriate. It would take too long here to describe this apparatus, which, moreover, is described in the third part of this work. I will therefore limit myself to giving an account of the results.

13.8 One or two hours after the substances are mixed together, especially if the temperature during the experiment is between 15 and 18 degrees, the first signs of fermentation appear. The liquid clouds up and becomes frothy. It releases bubbles which rise and burst on the surface. The quantity of bubbles quickly increases and very pure carbonic acid gas is rapidly released in great abundance, and this is accompanied by froth, which is only the yeast separating out. After several days, depending on the degree of heat, the motion

[146]   and the release of gas slow, though they do not stop entirely. And it is only after a rather long time that the fermentation is complete.

13.9 The weight of dry carbonic acid that is released in this operation is 35 livres, 5 ounces, 4 gros, 19 grains.

13.10 This gas, moreover, carries along with it a rather considerable portion of water which it holds in solution, and which is around 13 livres, 14 ounces, 5 gros.

13.11 There remains, in the vessel used for this procedure, a vinous liquid which is slightly acidic, at first cloudy, which then becomes clear, depositing a portion of the yeast. This liquid weighs, in total, 397 livres, 9 ounces, 29 grains.

13.7 *an apparatus*: The complete "pneumato-chemical apparatus" (3.27), shown here in a detail adapted from Plate X of Part III. The gaseous products pass first through a pair of separation vessels, then to Lavoisier's large pneumatic trough.

13.12 Finally, when all of the substances are separately analyzed and resolved into their respective constituent parts, the following results, after very painstaking labor, are obtained—results which will be gone over in detail in the memoirs of the Academy. [147]

Table of the Results Obtained by Fermentation

livres	ounces	gros	grains			livres	ounces	gros	grains
35	5	4	19	of carbonic acid composed of	oxygen	25	7	1	34
					carbon	9	14	2	57
408	15	5	14	of water composed of	oxygen	347	10		59
					hydrogen	61	5	4	27
57	11	1	58	of dry alcohol composed of	oxygen combined with hydrogen	31	6	1	64
					hydrogen combined with oxygen	5	8	5	3
					hydrogen combined with carbon	4		5	
					carbon	16	11	5	63
2	8			of dry acetous acid composed of	hydrogen		2	4	
					oxygen	1	11	4	
					carbon		10		
4	1	4	3	of residual sugar composed of	hydrogen		5	1	67
					oxygen	2	9	7	27
					carbon	1	2	2	53
1	6		50	of dry yeast composed of	hydrogen		2	2	41
					oxygen		13	1	14
					carbon		6	2	30
					azote			2	37
510						510			

13.12, Table. *oxygen combined with hydrogen, hydrogen combined with oxygen*: Why does Lavoisier distinguish between these two? The gaseous products of distillation were passed through a series of collection jars, two of which are illustrated in the apparatus shown opposite. Ideally, each jar would hold a single product; but in practice the separation would not have been complete. The individual entries represent the contents of separate jars, and Lavoisier's "combined" (*combiné*) gases are what we would now call a *mixture*.

[148]

Recapitulation of the Results Obtained by Fermentation

livres	ounces	gros	grains			livres	ounces	gros	grains
					from water	347	10		59
					from carbonic acid	25	7	1	34
409	10		54	of oxygen	from alcohol	31	6	1	64
					from acetous acid	1	11	4	
					from residual sugar	2	9	7	27
					from yeast		13	1	14
					from carbonic acid	9	14	2	57
					from alcohol	16	11	5	63
28	12	5	59	of carbon	from acetous acid		10		
					from residual sugar	1	2	2	53
					from yeast		6	2	30
					from water	61	5	4	27
					from water in alcohol	5	8	5	3
71	8	6	66	of hydrogen	combined with carbon in alcohol	4		5	
					from acetous acid		2	4	
					from residual sugar		5	1	67
					from yeast		2	2	41
		2	37	of azote				2	37

510   510

13.13 Although I have calculated these results to within a few grains, experiments like these by no means admit of such precision yet. But since I only worked with a few livres of sugar, and since I was obliged to adjust them to 100 livres in order to establish comparisons, I thought that I should retain the fractions as my calculations gave them.

[149]

13.14 Reflecting on the results that are presented in the tables above, it is easy to see clearly what occurs in vinous fermentation. Note first of all that, out of the 100 livres of sugar used, 4 livres, 1 ounce, 4 gros, 3 grains were not decomposed, so that only 95 livres, 14 ounces, 3 gros, 69 grains of sugar were actually used. That is to say, only 61 livres, 6 ounces, 45 grains of oxygen, 7 livres, 10 ounces, 6 gros, 6 grains of hydrogen, and 26 livres, 13 ounces, 5 gros, 19 grains of carbon were used. Now by comparing these quantities it will be seen that there is enough of these substances to form all of the spirit of wine (alcohol), all of the carbonic acid, and all of the acetous acid which was produced by the process of fermentation. It is therefore not at all necessary to assume that water is decomposed in this procedure, unless one claims that oxygen and hydrogen exist in the state of water in sugar. This is something I do not believe, since I have established, on the contrary, that in general the three constitutive principles of vegetables (hydrogen, oxygen, and carbon) are in a state of equilibrium; that this state of equilibrium persists so long as it is not disturbed, whether by a change in temperature or by some double [150] affinity; and that only then, by combining two by two, will these principles form water and carbonic acid.

13.15 Thus the only effect of vinous fermentation is to separate the sugar, which is an oxide, into two parts. By oxygenating the first at the expense of the second, carbonic acid is formed; by deoxygenating the second in favor of the first, a combustible substance, alcohol, is formed. So if it were possible to recombine these two substances, alcohol and carbonic acid, sugar would form again. It is noteworthy, moreover, that in alcohol, hydrogen and carbon are not in the state of an oil. They are combined with a portion of oxygen which makes them miscible with water. The three principles, oxygen, hydrogen, and carbon, are therefore also here in a kind of equilibrium; and, in fact, by passing them through a red hot glass or porcelain tube, they are recombined in groups of two; and water, hydrogen, carbonic acid, and carbon are recovered.

13.14 *comparing these quantities*: Expressed in decimal form, 95.905 livres of sugar were consumed. The products comprise 57.702 livres alcohol, 35.346 livres carbonic acid, and 2.5 livres acetous acid, totalling 95.548 livres—nearly equal to the initial quantity of sugar. The constituents of sugar are thus *sufficient* to form all the substances that were produced in the fermentation process. But did sugar *actually become* these substances? Lavoisier will make that very claim in the next paragraph.

13.15 *Thus the only effect of vinous fermentation is to separate the sugar ... into two parts*: But the previous paragraph identified *three* products, not two. Of these, Lavoisier proceeds to recognize only *alcohol* and *carbonic acid*; why does he ignore the third product, acetous acid? We will learn in Chapter Fifteen that he regards acetous acid (vinegar) as a secondary product, arising when some of the principal products, alcohol and carbonic acid, combine with additional oxygen (15.5).

Lavoisier does not comment on the discrepancy between the initial quantity of yeast (2.761 livres) and the amount recovered (1.380 livres). Was yeast consumed? That would contradict the impression created in paragraph 13.6, that yeast takes no part in the ongoing fermentation process.

13.16 In my first memoirs on the formation of water, I explicitly advanced the claim that this substance, formerly regarded as an element, decomposes in many chemical operations, notably in vinous fermentation. At that time I assumed that water existed fully formed in the sugar. But now I am persuaded that the sugar only contains the materials suitable for forming it. Obviously it must have cost me a great deal to abandon my first ideas. It was only after several years of reflections, and a long series of experiments and observations on vegetables, that I came to my current position. [151]

13.17 I will conclude what I have to say on vinous fermentation by observing that it provides a way of analyzing sugar and, in general, any vegetable substance that can be fermented. Indeed, as I already indicated at the beginning of this chapter, I can consider the substances subjected to fermentation and the results obtained after fermentation as an algebraic equation. And by successively assuming that each term in this equation is unknown, I can derive a value for it, and in this way correct both the experiment through calculation and the calculation through experiment. I have often taken advantage of this method to correct the first results of my experiments, and to help me determine which precautions I needed to take when I performed them again. But this is not the moment to go into these details, about which, moreover, I have already expanded at great length in the memoir on vinous fermentation which I presented at the Academy and which will be published very shortly. [152]

13.17 *an algebraic equation*: Recall Lavoisier's use of equation formalism in (13.4), where, however, its significance seemed primarily rhetorical. But he is about to claim that the equation form serves a more directly functional purpose.

by successively assuming that each term in this equation is unknown, I can derive a value for it: This is precisely how one solves a multivariable equation, provided one has provisional values for the "known" variables. The method of choosing each quantity in turn as "unknown" is a recursive procedure which reconciles measured and calculated values as nearly as possible; that is what Lavoisier means by correcting "the experiment through calculation and the calculation through experiment." Did Lavoisier actually use this method to construct the tables in this and other chapters? His experimental records do not unambiguously answer that question; see Frederick Holmes, *Lavoisier and the Chemistry of Life, An Exploration of Scientific Creativity* (University of Wisconsin Press, 1987), Chap. 12.

14.1 I just showed how sugary bodies were decomposed when diluted with a certain amount of water and assisted by a gentle heat. I showed how the three principles (oxygen, hydrogen, and carbon) were in a state of equilibrium while they were constituents of sugar and so formed neither water, nor oil, nor carbonic acid, and then were separated so as to be combined in a new arrangement. And I also showed how one portion of carbon reunited with oxygen to form carbonic acid, and how another portion of carbon combined with hydrogen and water to form alcohol.

14.2 The phenomena of putrefaction occur in a similar way, by virtue of some very complex affinities. In this process, the three constitutive principles of the bodies also cease to remain in a state of equilibrium. Instead of a ternary compound, they form binary compounds, but the result of these combinations is quite different from that of vinous fermentation. In the lat- [154] ter, some of the principles forming the vegetable substance, hydrogen, for example, remain united to a portion of water and of carbon to form alcohol. In putrid fermentation, on the contrary, the entire quantity of hydrogen dissipates in the form of hydrogen gas. At the same time, oxygen and carbon, reuniting with caloric, escape in the form of carbonic acid gas. Finally, when the operation is entirely completed, especially if the quantity of water necessary for putrefaction has not been found lacking, there remains nothing more than vegetable earth mixed with a little carbon and iron.

14.3 The putrefaction of vegetables is therefore nothing else than a complete analysis of the vegetable substances in which all of their constitutive principles are released in gaseous form, with the exception of earth, which remains in the state which we call "compost."

14.4 In the third part of this work I will give the reader some idea of the equipment which can be used in experiments like these.

14.5 Such is the result of putrefaction when the body submitted for the process contains only oxygen, hydrogen, carbon, and a little earth. But this is a rare case; and it even appears that when these substances are alone, they ferment with difficulty: they ferment poorly, and a considerable amount of [155] time is required for them to putrefy completely. It is not the same when the substance undergoing fermentation contains some azote, which is true for all animal matter and fairly often for vegetable matter. This new ingredient marvelously favors putrefaction; and this is why animal matter is mixed with vegetable matter when one wishes to speed up the process of putrefaction. And in this mixing consists nearly the entire science of fertilizers and manure.

14.6 But adding azote to putrefying materials does not only accelerate the process of putrefaction. Combining with hydrogen, azote forms a new substance known by the name of "volatile alkali" or "ammoniac." The results obtained through analysis of animal matter by different processes leaves no doubt about the nature of the principles which constitute ammoniac. In each instance where azote is previously separated from these materials, they no longer yield ammoniac, and they yield it only insofar as they contain azote. [156] This composition of ammoniac is also confirmed by analytical experiments related in detail by Berthollet.* He has given different ways of decomposing this substance and of obtaining separately the two principles, azote and hydrogen, that compose it.

14.7 I already asserted above (see Chapter Ten) that nearly all of the combustible bodies were capable of combining with one another. Hydrogen gas very clearly possesses this property. It dissolves carbon, sulphur, and phosphorus, and from these combinations there results what I have named in that chapter carbureted hydrogen gas, sulphureted hydrogen gas, and phosphorated hydrogen gas. The two latter gases have a peculiarly disagreeable odor. That of sulphureted hydrogen gas is very much like that of rotten and spoiled eggs; that of phosphorated hydrogen gas is absolutely the same as that of rotten fish. Finally, ammoniac has an odor which is neither less penetrating, nor less disagreeable than the others. The combination of these different odors results in the odor that animal materials in the process of fermentation emit, which is so fetid. At times, the odor of ammoniac is predominant, being easily recognized in that it stings the eyes. At times the odor [157] of sulphur predominates, as in the case of fecal matter, and finally, at times it is the odor of phosphorus, as in rotten herring.

14.8 Up until now, I have supposed that nothing could interfere with the course of fermentation, or interrupt its effects. But de Fourcroy and Thouret have observed certain peculiar phenomena with respect to corpses buried

* *Mémoires de l'Académie* for 1785, p. 316

14.5 *true for all animal matter and fairly often for vegetable matter*: Recall Lavoisier's remarks on azote in vegetable and animal matter in paragraphs 12.10 and 12.11.

at a certain depth, and safeguarded up to a certain point from contact with air. They have commented that often the muscular part of the corpse is converted into true animal fat. This phenomenon depends on this, that in some peculiar way the azote contained by these animal materials would already have been released, and that all that remains is hydrogen and carbon, that is to say, the proper materials for producing fat. This observation on the possibility of converting animal materials into fat may one day lead to important discoveries concerning the advantages that can be drawn from them for the use of society. Animal excrement, such as fecal matter, is primarily composed of carbon and hydrogen. It is therefore very close to the state of an oil, and, in fact, furnishes a great amount of this by distillation over an open fire. But the [158] unbearable odor which accompanies all products extracted from it does not allow us to hope, at least for a long time, that we may use it for anything other than the production of manure.

14.9 Because the composition of animal matter is not yet known precisely, I have given only glimpses of ideas in this chapter. Animal matter is known to be composed of hydrogen, carbon, azote, phosphorus, and sulphur, and is brought to the state of an oxide by a greater or lesser quantity of oxygen. But we are absolutely ignorant as to the proportion of these principles. Time will bring to completion this part of chemical analysis, as it has already completed several others.

Chapter Fifteen
On Acetous Fermentation

15.1 Acetous fermentation is nothing else than the acidification of wine, which occurs in open air by the absorption of oxygen. The resulting acid is acetous acid, commonly called vinegar. It is composed of hydrogen and carbon, combined together in a proportion not yet determined, and brought to the state of an acid by oxygen.

15.2 Since vinegar is an acid, analogy alone was leading us to conclude that it contains oxygen. But this truth is further demonstrated by direct experiments. In the first place, wine can be converted into vinegar only through contact with the air, and only insofar as this air contains oxygen gas. Second, this process is accompanied by a decrease in the volume of the air in which it is performed, and this decrease in volume is caused by the absorption of oxygen gas. Third, wine can be transformed into vinegar by any other means of oxygenation.

15.3 Apart from these facts which prove that acetous acid is the result of the oxygenation of wine, an experiment by Chaptal, professor of chemistry at Montpellier, illustrates clearly what happens in this process. He takes carbonic acid gas which is released from fermenting beer. He absorbs it with water until the point of saturation, that is, until the water has absorbed a quantity nearly equal to its volume. He places this water in a cellar in vessels communicating with the air, and after a certain amount of time the whole is converted into acetous acid. Carbonic acid gas from vats of fermenting beer is not entirely pure. It is mixed with a little alcohol which remains in solution.

15.1 *acetous acid, commonly called vinegar*: The French adjective *aceteaux* (acetous) denoted the sourness characteristic of vinegar even before Lavoisier's time; of course the adjectival suffix did not then have the systematic significance that Lavoisier would later give to it (6.3). In addition to ordinary vinegar, both medicine and chemistry recognized "radical vinegar," a related but apparently different substance, which Lavoisier names "acetic acid" in Part III of the complete Treatise. Ordinary vinegar is a dilute acid solution; radical vinegar was eventually recognized as the identical acid in dry crystalline ("glacial") form. Lavoisier's term "acetous acid" has therefore been abandoned in chemistry; and the acid of vinegar is now known as *acetic acid*.

Thus in the water impregnated with carbonic acid released from vinous fermentation, there exist all the materials required to form acetous acid. The alcohol furnishes hydrogen and a portion of the carbon. Carbonic acid furnishes carbon and oxygen. Lastly, atmospheric air must furnish whatever amount of oxygen is lacking to bring the mixture to the state of acetous acid.

15.4 By this it is shown that it is only necessary to add hydrogen to the carbonic acid in order to constitute it as acetous acid, or to speak more generally, in order to transform it into a given vegetable acid, in accordance with the degree of oxygenation; that, contrariwise, it is only necessary to subtract   [161] hydrogen from vegetable acids in order to convert them into carbonic acid.

15.5 I will expand no further on the subject of acetous fermentation, in regard to which we do not yet have exact experiments. While the principal facts are known, numerical exactitude is lacking. Besides, it can be seen that the theory of acetification is closely tied to that of the constitution of all vegetable acids and oxides, and we do not yet know the proportion of the principles of which they are composed. It is easy, however, to perceive that this whole branch of chemistry, like every branch of chemistry, is progressing rapidly towards its perfection, and that it is much simpler than previously thought.

15.4 *in accordance with the degree of oxygenation*: Recall that in Lavoisier's system, a single acifiable base may give rise to multiple acids by undergoing multiple degrees of oxygenation (6.3, 6.10 and comment).

Chapter Sixteen

*On the Formation of Neutral Salts, and the Different Bases
that Enter into their Constitution*

16.1 We have seen how a small number of simple substances, or at least those which up until now have not been decomposed, such as azote, sulphur, phosphorus, carbon, the muriatic radical, and hydrogen, by combining with oxygen formed all of the oxides and acids in the vegetable and animal kingdoms. We have admired the simplicity of means by which nature multiplied their properties and forms, whether by combining together as many as three or four acidifiable bases in different proportions, or by varying the amount of oxygen required to acidify them. We shall find her no less varied, no less simple, and, above all, no less abundant in the order of things discussed below.

16.2 Acidifiable substances, when they combine with oxygen and are converted into acids, acquire a strong tendency to form compounds. They acquire the capacity to unite with earthy and metallic substances, and this union

produces the neutral salts. Acids can therefore be regarded as true salifying principles, and the substances to which they are united so as to form neutral salts can be regarded as salifiable bases. It is exclusively the combination of

---

16.2 *neutral salts*: Dissolved in water, a small number of salts produce acidic solutions; others salts yield solutions having properties resembling those of soda, potash and ammoniac—the three "alkalis" (16.15, comment). The majority of known salts do neither, and these are the salts Lavoisier calls "neutral." To a modern reader, "neutral" may imply *balance between opposites*; but, as we are about to discover, Lavoisier's understanding of salification contains no suggestion of antagonism or conflict. What, then, does "neutral" mean for Lavoisier?

salifying ... salifiable: Note the contrast in these epithets between *active* (-ing) and *potential* (-able) forms. Far from being sums of opposites, salts arise through the salifying principles with the salifiable bases that will concern us in this chapter.

16.3 This way of viewing acids does not allow me to consider them as salts, even though they possess some of the main properties of salts, such as solubility in water, etc. Acids, as I have already pointed out, result from first-order combinations. They are formed when two principles which are simple (or at least act as if they were simple) unite. And for this reason they are classified as "mixts," to use Stahl's term. Neutral salts, on the other hand, are combinations of a different order. They are formed by joining two "mixts," and so they fall under the class of "composites." For the same reason, I will no longer assign alkalis or earthy substances such as lime, magnesia, etc., to the class of salts.* I will designate by the name "salt" only the composites formed by joining a simple oxygenated substance to a given base.

[164]

16.4 In the preceding chapters, I have sufficiently expanded on the formation of acids, and so I will add nothing further here. However, I have not yet said anything about the bases that can combine with them to form neutral salts. These bases, which I call "salifiable," are as follows:

> Potash
> Soda
> Ammoniac
> Lime
> Magnesia
> Barytes
> Alumina
> All metallic substances

---

* Perhaps it will be considered a defect in the method I have adopted that it compels me to reject the alkalis from the class of salts. I agree that this is a reproach that can be made. But this inconvenience is compensated by advantages so great that I did not believe it ought to deter me.

---

the transformation of a passive subject by an active power. In retrospect, we can see that they represent the pinnacle of Lavoisier's theme of *oxidation*, one that began with oxides and acids (Chapter Five) and culminates with salts in the present chapter and the next—the final two chapters of Part I.

But what leads Lavoisier to believe that *acids* are "true salifying principles"—that is, the active components? Since *oxidation* has been the ongoing motif, perhaps acids are implicated insofar as it is acids which carry the oxygen, a fact Lavoisier emphasizes in the opening sentence of this paragraph.

16.5 I will say a word concerning the origin and nature of each one of these bases.

On Potash

[165]

16.6 We have already pointed out that whenever a vegetable substance is heated in a distilling apparatus, the principles composing it, oxygen, hydrogen, and carbon, which form a ternary compound in a state of equilibrium, reunite in pairs in accordance with the affinities they have for one another at different temperatures. Thus, at first contact with the fire, and as soon as the heat exceeds the temperature of boiling water, oxygen and hydrogen unite to form water. Soon afterwards, some of the carbon and some of the hydrogen combine to form oil. Then, once the distillation further progresses and substances become red hot, the oil and even the water which had formed are decomposed: oxygen and carbon form carbonic acid, and a great quantity of hydrogen gas breaks free, is released and escapes. In the end, only carbon remains in the retort.

16.7 Most of these phenomena are met with again whenever vegetable matter burns in the open air. But then the presence of the air introduces three new ingredients, and at least two of these change the results of the process considerably. These ingredients are oxygen from the air, azote, and caloric. As the

---

16.6 *We have already pointed out*: Lavoisier outlined the distillation of vegetable matter in (12.7). But his present review of that process differs from the earlier account in one particular. There he reported that distillation at temperatures near that of boiling water left a residue of carbon, while distillation at red heat left no carbon behind. In this paragraph, however, he maintains that "carbon remains in the retort" after distillation, even at red heat. The discrepancy is confusing, but the following comments will propose a resolution.

oxygen and hydrogen unite to form water: The distillation products will depend on *how much* of this water escapes in the form of vapor.

once the ... substances become red hot, the oil and even the water which had formed are decomposed: If not too much water is lost to vapor, decomposition of the remaining water will supply enough oxygen to convert all the carbon to carbonic acid, leaving no residue—the case in (12.7). Otherwise, however, some of the carbon "remains in the retort," as Lavoisier reports here. The process is quite sensitive to temperature; and since "red heat" actually covers a wide range of temperatures, both reported outcomes are possible.

16.7 *vegetable matter burns in the open air*: Burning, not distillation, is the actual focus of Lavoisier's interest; he brought up the case of distillation (wherein oxygen is limited) merely to contrast it with combustion, in which oxygen is plentiful.

hydrogen (whether from the vegetable matter or from the water as it decomposes) is expelled by the fire's progress in the form of hydrogen gas, it ignites as soon as it is in contact with air. It then becomes water again, and the caloric freed from the two gases, at least the greatest part of it, produces a flame.

16.8 Then, when all of the hydrogen gas has been expelled, burned, and reduced to water, the carbon which remains burns as well, though without flame. It forms carbonic acid, which escapes, carrying with it a share of caloric which establishes it as a gas; the remaining caloric is freed, escapes, and produces the heat and light that we observe in the combustion of charcoal. And so all of the vegetable matter is reduced to water and carbonic acid. There remains only a small portion of a gray, earthy matter, known by the name of ash, and which contains the only truly fixed principles that enter into the constitution of vegetables.

16.9 This earth or ash, whose weight does not usually exceed one-twentieth of the vegetable matter, contains a substance of a particular kind, known under the name of fixed vegetable alkali, or potash.

16.10 To obtain it, one pours water over the ashes; the water fills with potash which is soluble, leaving behind the ashes which are insoluble. Then by evaporating the water, one obtains potash that is fixed, even at a great degree of heat, and that remains in a white and solid form. My object here is certainly not to describe the art of preparing potash, even less the means by which to obtain it pure. I include these details here simply in accordance with the rule that I made for myself of introducing no word that has not been defined. [167]

16.11 The potash that is obtained by this procedure is always more or less saturated with carbonic acid, and the reason for this is easy to grasp: since

---

16.8 *the carbon which remains ... forms carbonic acid, which escapes*: Thereby leaving no residue of carbon.

There remains only a small portion of ... ash: This is unexpected! Lavoisier said nothing about "ash" in (12.7), even in cases that left no carbon behind. On the contrary, he rather strongly implied that the entirety of the distilled matter passed over in gaseous form.

16.9 *This earth or ash ... contains a substance ... known under the name of fixed vegetable alkali, or potash*: The name *potash* (pot ash) is almost self-explanatory. Notice that it is an instance of synecdoche—naming a part by the whole.

The alternate name, "fixed vegetable alkali," is more consequential. In it the noun, *alkali*, is modified by two adjectives. The adjectives are easy to interpret: the substance is *fixed*, that is, not volatile, as we will learn shortly; and it has a *vegetable* source, as we already know. But what is an *alkali*? We will learn more in (16.15).

potash is only formed, or at least is only freed, to the extent that carbon from the vegetable matter is converted into carbonic acid by the addition of oxygen from air or from water, it follows that each particle of potash is found, at the moment of its formation, in contact with a particle of carbonic acid. And since there is a great affinity between these two substances, they must combine. Although of all the acids, carbonic acid clings to potash the least, it is nevertheless difficult to separate the last portions of it. The most commonly

[168]   used method consists in dissolving potash in water, adding two or three times its weight of quicklime, and filtering and evaporating it in closed vessels. The saline substance obtained is potash, nearly completely divested of carbonic acid.

16.12 In this state, the potash is not only soluble in water in at least equal amounts, but it even attracts the water in the air, and with astonishing eagerness. As a consequence, it provides a means of drying air or gases to which it is exposed. It is equally soluble in spirit of wine (alcohol), except when it is saturated with carbonic acid, which is not soluble in this solvent. This circumstance furnished Berthollet with a way to attain perfectly pure potash.

16.13 All vegetables produce some amount of potash when incinerated. But it is not possible to obtain potash of equal purity from every vegetable: it is usually mixed with different salts from which it is easily separated.

16.14 One can hardly doubt that the ashes, or in other words the earth that vegetable matter leaves behind when burned, were already present before combustion. This earth forms, it seems, the bones or skeleton of the plant.

[169]   But this is not true for potash; this substance has not yet been separated

---

16.14 *One can hardly doubt that the ashes … were already present before combustion*: The ashes' ability to survive combustion is reason to believe that they took no part in any process of composition or decomposition; and if so, the ash components were probably united firmly in the vegetable matter from the start. Notice that the question, whether a given substance existed already prior to the process that yielded it or was brought into existence by that process, has become almost a canonical one for Lavoisier; he raised it previously at (12.6) and (13.16) and will do so again.

the bones or skeleton of the plant: Animal bones are notably resistant to burning.

but this is not true for potash: In contrast to ashes, we *can* doubt that potash pre-existed in the vegetable matter. The whitish solid obtained by evaporating ash-water (16.10) is not a simple substance but contains carbonic acid (16.11). Although this solid itself is sometimes (and confusingly) called potash, that name ought to be reserved for the substance which remains after the carbonic acid is removed.

from vegetable matter, except by means of processes or intermediaries that can provide oxygen or azote, such as combustion or combination with nitric acid. So it has by no means been demonstrated that this substance is not a product of these processes. I have begun a series of experiments to this end, of which I will soon be ready to give an account.

On Soda

16.15 Like potash, soda is an alkali that is extracted from the ashes of plants through lixiviation; but only from those plants that grow by the seashore, and primarily from kali, from whence the name "alkali" was given to it by the Arabs. While it has several properties in common with potash, it has others which distinguish the two. In general, both of these two substances have peculiar properties which they bring to their respective saline compounds. Soda, as obtained by the lixiviation of marine plants, is most often entirely saturated with carbonic acid, but it does not attract moisture from the air like potash. On the contrary, it dries out in the air and its crystals effloresce and convert into a white powder. This powder has all of the properties of soda and differs from it only insofar as it has lost its water of crystallization. [170]

16.16 At the present time, we are as far from knowing which principles compose soda as we are from knowing which principles compose potash. It is not even certain whether soda already exists in vegetables fully formed prior to combustion. Analogy might lead us to believe that azote is one of the constituent principles of alkalis in general; and we have proof of this with respect to ammoniac, as I am going to show. But with respect to potash and

---

[potash] has not yet been separated from vegetable matter except by means ... such as combustion or combination with nitric acid: Combustion supplies oxygen; nitric acid supplies azote. If these substances contributed to the potash so obtained, the potash will have been "a product of these processes" rather than a pre-existing constituent of vegetable matter.

16.15 *Like potash, soda is an alkali*: Lavoisier is about to explain the origin of the name "alkali." The name "soda" (French *soude*) has a similar origin but has come to denote a particular substance, while "alkali" names a class of substances.

lixiviation: As described in (16.10), the process of dissolving a substance in water, then evaporating the water to extract the soluble portion.

primarily from kali, from whence the name "alkali": *Al* is the Arabic definite article; *kali*, or glasswort, is a common seaside herb. Thus soda is the paradigmatic alkali, from whose properties the whole class of alkalis was named. Alkalis produce a burning sensation on the tongue, among other properties; in 18th century English, "alkaline" was a common epithet denoting something bitter, caustic, and soapy.

proof of this with respect to ammoniac: That is, proof that ammoniac contains azote.

soda, there are only superficial conjectures that have not yet received confirmation from any decisive experiment.

On Ammoniac

16.17 Since we had no precise knowledge to present concerning the composition of soda and potash, we were obliged to restrict ourselves in the preceding two sections to indicating the substances from which they are extracted, and the means employed to obtain them. It is otherwise with ammoniac, which our predecessors called "volatile alkali." Berthollet, in a published memoir,* successfully proved by means of decomposition that 1000 parts by weight of this substance were composed of approximately 807 parts of azote and 193 parts of hydrogen.

[171]

16.18 This substance is obtained primarily by distilling animal matter. The azote, a constituent principle of animal matter, unites with hydrogen in the proper proportion for this compound, and ammoniac is formed. But pure ammoniac is not obtained by this method; it is mixed with water, oil, and very often with carbonic acid. In order to separate it from all of these substances, it is combined first with an acid such as muriatic acid. Then it is released from the acid, whether by an addition of lime or by an addition of potash.

16.19 When ammoniac has in this way been brought to its greatest degree of purity, it can only exist in a gaseous form at the temperature in which we live. It has an exceedingly pungent odor. Water absorbs great quantities of it, especially if the ammoniac is cold and if it is not only chilled but compressed. Water saturated with ammoniac in this way has been called "volatile alkali fluor." We will call it simply ammoniac, or liquid ammoniac, and we will name the same substance, when in an aeriform state, "ammoniacal gas."

[172]

On Lime, Magnesia, Barytes, and Alumina

16.20 The composition of these four earths is completely unknown. And since we have not yet succeeded in determining their constituent and elementary parts, we may consider them simple substances while we await new discoveries. Art has, then, no role in forming these earths; nature presents them to us completely formed. But as they have for the most part, especially the first three, a great tendency to combine, they are never found in isolation.

* *Recueil de l'Académie*, 1784, p. 316.

16.17 *volatile alkali*: As we gather from the present chapter, there are three traditionally recognized "alkalis": *fixed vegetable alkali* (potash), *alkali* proper (soda) and *volatile alkali*, henceforth to be styled *ammoniac*.

Lime is nearly always saturated with carbonic acid, and in this state it forms chalk, calcareous spars, some varieties of marble, etc. It is sometimes saturated with sulphuric acid, as in gypsum and plaster stones; at other times when saturated with fluoric acid it forms vitreous or fluor spars. And finally, the waters of the sea and saline springs contain it combined with muriatic acid. Of all the salifiable bases, lime is the most abundant and ubiquitous in nature.   [173]

16.21 Magnesia is often found in mineral water, where it is most commonly combined with sulphuric acid. There is an abundance of it in sea water, where it is combined with muriatic acid. Finally, it is a constituent of many stones.

16.22 Barytes is much less prevalent than the two preceding earths. It is found, in the mineral kingdom, combined with sulphuric acid where it then forms heavy spars. Sometimes, though less often, it is combined with carbonic acid.

16.23 Alumina, or the base of alum, has less of a tendency to form compounds than the preceding bases. In addition, it is often found in the state of alumina, uncombined with any acid. It is mainly found with clays, of which, properly speaking, it constitutes the base.

On Metallic Substances

16.24 With the exception of gold and sometimes silver, the metals rarely show themselves in the mineral kingdom in their metallic form. They are usually more or less saturated with oxygen, or combined with sulphur, arsenic, sulphuric acid, muriatic acid, carbonic acid, or phosphoric acid.   [174] Docimasy and metallurgy provide instructions for separating them from all foreign substances, and we refer the reader to works devoted to this aspect of chemistry.

16.25 It is likely that we are aware of only a part of the metallic substances existing in nature. For example, all those that have a greater affinity for oxygen than for carbon cannot be reduced or restored to a metallic state. Consequently they can present themselves to our eyes only in the form of oxides, which we mistake for earths. It is very likely that barytes, which we have just classified among the earths, is in this category. For under the

---

16.23 *alum*: A number of substances are called "alum," but the alum being discussed here is an astringent mineral salt, widely employed as a dye fixative among other uses.

16.25 *which we mistake for earths*: That is, we suppose them to be irreducible substances—as earths were generally presumed to be (16.20).

particular circumstances of some experiments, it displays characteristics that give it a very close resemblance to a metal. In the end, it could turn out that all those substances to which we give the name "earths" are only metallic oxides that are irreducible by the means we currently employ.

16.26 Whatever the case may be, the metallic substances that we know, [175] those that we are able to obtain in a metallic state, are the following seventeen:

Arsenic	Iron
Molybdenum	Tin
Tungsten	Lead
Manganese	Copper
Nickel	Mercury
Cobalt	Silver
Bismuth	Platinum
Antimony	Gold
Zinc	

16.27 Here, I will consider these metals only as salifiable bases, and I will not go into any detail about their properties insofar as they relate to the arts and uses of society. From this point of view, each metal would require a complete treatise, and I would completely go beyond the bounds which I have prescribed for myself.

16.25, cont. *it could turn out that ... "earths" are only metallic oxides*: Three of the "earths" named in paragraphs (16.20–16.23) have indeed proved to be oxides; but *barytes*, which Lavoisier has just singled out as a particularly likely candidate, is a sulphate.

Chapter Seventeen

Continuation of Reflections on the Salifiable Bases, and on the Formation of
Neutral Salts

17.1 So much, then, for the salifiable bases, i.e., bases which can combine with acids to form neutral salts. But it is necessary to observe that alkalis and earths enter into the composition of neutral salts in a pure and simple way, without any intermediary to unite them. Metals, on the contrary, can combine with acids only to the extent that they have been previously oxygenated to a greater or lesser degree. Thus, strictly speaking, metals are not at all soluble in acids: only the metallic oxides are. Thus when one places a metallic substance in an acid, the first condition that must be met for it to dissolve is that it must first oxidize there. And it can do so only by removing oxygen, either from the acid or from the water in which this acid is diluted. In other words, a metallic substance can dissolve in an acid only to the extent that the oxygen that enters into the composition either of the water or of the acid [177] has a greater affinity for the metal than it does for hydrogen or the acidifiable base. Or, what amounts to the same thing, metallic dissolution occurs only insofar as the water or the acid decomposes.

17.2 This simple observation, that escaped even the illustrious Bergman, explains the principal phenomena associated with dissolving metals. The first and most striking of these is the phenomenon of effervescence, or, to speak in a less equivocal way, the release of gas which takes place during dissolution. When nitric acid is used as a solvent, this gas is nitrous gas. When sulphuric acid is used as a solvent, it is either sulphurous acid gas or hydrogen gas, according to whether the sulphuric acid or the water is responsible for oxidizing the metal.

17.3 Since nitric acid and water are composed of substances that can only exist as gases at the temperature in which we live, it is apparent that as soon as

---

17.1 *metals ... can combine with acids only to the extent that they have been previously oxygenated*: Lavoisier will explain the grounds for this statement in the following paragraphs.

effervescence, or ... the release of gas: Some metals, notably zinc, dissolve rapidly in acids (7.9, comment) and release a profusion of bubbles; this is effervescence. But gas is released even when the dissolution is slow.

the oxygen is removed, the principles to which it was attached ought immediately to expand and assume gaseous form. This rapid transition from the
liquid to the gaseous state constitutes effervescence. It is the same in the case

[178]   of sulphuric acid. In general, metals never remove all of the oxygen from this
acid, especially by the humid way. They never reduce it to sulphur, but rather
to sulphurous acid which also can only exist in a gaseous state at the degree
of temperature and pressure in which we live. This acid must, for this reason,
be released in the form of a gas; and again, when the gas is released, effervescence occurs.

17.4 A second phenomenon is that all metallic substances dissolve without effervescence in acids when they have been oxidized before dissolution.
It is clear in this case that the metal, as it no longer has to be oxidized, no
longer tends to decompose either the acid or the water. Therefore, there
can no longer be any effervescence, since the effect that was producing it
no longer occurs.

17.5 A third phenomenon is that all metals dissolve without effervescence
in oxygenated muriatic acid. What happens in this reaction deserves several
special observations. In this case, the metal removes the excess oxygen from
the oxygenated muriatic acid, creating, on the one side, metallic oxide and,
on the other, ordinary muriatic acid. If there is no effervescence in dissolu-
tions like these, it is not because muriatic acid cannot naturally exist in the

[179]   form of a gas at the degree of temperature in which we live, but rather this
gas finds in oxygenated muriatic acid more water than is necessary for it to
be retained and remain in liquid form. So it is not released as in the case of
sulphurous acid, and after having instantly combined with the water, it then
peacefully combines with the metallic oxide, which it dissolves.

17.3 *the humid way*: That is, in the presence of water.

17.5 *metals dissolve without effervescence in oxygenated muriatic acid*: This
acid, then, constitutes an exception, which Lavoisier attempts to explain in the
remainder of this paragraph. But recall that muriatic acid, whether "oxygenated"
or not, has already shown itself to be an anomaly among acids in several other
ways as well (6.10).

*the metal removes the excess oxygen from the oxygenated muriatic acid, creating ...
muriatic acid*: This of course comports with Lavoisier's general view that acids
having the same radical differ from one another in respect of their degree of
oxygenation. But muriatic acid *does* exist as a gas under ordinary conditions (6.7);
why then does it not escape in effervescence as other acid gases do?

*this gas finds in oxygenated muriatic acid more water than is necessary for it to ...
remain in liquid form*: This is a little misleading. The water in question is not "in"
oxygenated muriatic acid proper but is, rather, the medium in which that acid

17.6 A fourth phenomenon is that the metals that have little affinity for oxygen, and cannot attract this principle with enough strength to decompose either the acid or the water, are absolutely insoluble. It is for this reason that silver, mercury, and lead are insoluble in muriatic acid when they are introduced to this acid in their metallic state. But if they are previously oxidized, in any manner whatsoever, they immediately become very soluble, and dissolve without effervescence.

17.7 Oxygen is thus the link that unites metals and acids, and since this holds for all metals and all acids, we might be led to suppose that all substances which have a great affinity for acids contain oxygen. It is therefore fairly likely that the four salifiable earths we have indicated above contain [180] oxygen, and that it is through this common *latus* that they unite with the acids. This consideration would seem to support what I previously put forward in the section on earths, namely, that these substances might be nothing but oxidized metals which cannot be reduced for the reason that oxygen has more affinity for them than for charcoal. Nevertheless, this remains a mere conjecture which can only be confirmed or refuted by later experiments.

17.8 The following are the acids which are known at the present time. As we designate them, we will indicate the name of the radical or acidifiable base of which they are composed.

Names of the acids	Names of the acidifiable base or radical of each acid, with some observations
1. Sulphurous 2. Sulphuric	Sulphur
3. Phosphorous 4. Phosphoric	Phosphorus
5. Muriatic 6. Oxygenated Muriatic	Muriatic radical
7. Nitrous 8. Nitric 9. Oxygenated Nitric	Azote
10. Carbonic	Carbon

---

is dissolved. Therefore when *muriatic acid* is produced, it too dissolves in the water—as it has a great propensity to do (6.7).

17.7 *latus*: Latin *latus* literally means "side." Here Lavoisier gives it a sense resembling "aspect" or, as in the initial sentence of this paragraph, "link."

earths ... might be nothing but oxidized metals which cannot be reduced: Lavoisier put forth this conjecture previously in paragraph 16.25.

[181]

11. Acetous

12. Acetic

13. Oxalic

14. Tartarous

15. Pyro-tartarous

16. Citric

17. Malic

18. Pyro-lignous

19. Pyro-mucous

All these acids appear to be formed by the combination of two acidifiable bases, carbon and hydrogen, and differ among themselves only by the difference in ratio between these two bases and the oxygen which acidifies them. But there is not yet a series of well-designed experiments on this matter.

20. Gallic

21. Prussic

22. Benzoic

23. Succinic

24. Camphoric

25. Lactic

26. Saccho-Lactic

Knowledge of the nature of the radicals of these acids is still quite imperfect. All that is known is that carbon and hydrogen are the principal parts, and prussic acid contains azote.

27. Bombic

28. Formic

29. Sebacic

These acids and all those obtained by oxygenating animal matter, appear to have carbon, hydrogen, phosphorus, and azote for their acidifiable base.

30. Boracic	Boracic radical	The nature of these two radicals is completely unknown.
31. Fluoric	Fluoric radical	
32. Antimonic	Antimony	
33. Argentic	Silver	
34. Arsenic	Arsenic	
35. Bismuthic	Bismuth	
36. Cobaltic	Cobalt	
37. Cupric	Copper	
38. Stannic	Tin	
39. Ferric	Iron	
40. Manganic	Manganese	
41. Hydrargiric	Mercury	
42. Molybdic	Molybdenum	
43. Nickelic	Nickel	
44. Auric	Gold	
45. Platinic	Platinum	
46. Plumbic	Lead	
47. Tungstic	Tungsten	
48. Zincic	Zinc	

17.9 Clearly, then, the number of acids is forty-eight, including the seven- [182]
teen metallic acids which are still little known, but about which Berthollet
will soon publish an important work. We certainly cannot yet boast that we
have discovered them all. But on the other hand, it is likely that a deeper
examination will reveal that some of the vegetable acids now considered distinct are of the same kind. For the rest, the table given here can only represent
chemistry in its current state. We can do nothing more but provide principles
for naming those bodies which will be discovered in the future, in conformity with the self-same system.

17.10 The number of salifiable bases, that is, those capable of being converted into neutral salts by acids, is twenty-four, which are the following:
three alkalis, four earths, and seventeen metallic substances.

17.11 The total number of neutral salts that can be conceived in the present
state of our knowledge is therefore 1,152. But this assumes that metallic acids
can dissolve other metals. And this solubility of metals that oxygenate one
another is a new science that has not yet even been broached: it is upon this [183]
part of science that all vitreous metallic compounds depend. Furthermore,
it is likely that not all of the conceivable saline compounds are possible,
which must considerably reduce the number of salts that nature and art can
form. But even if we were to assume only five or six hundred possible kinds
of salts, it is clear that if we wished to give arbitrary names to all of them in
the manner of earlier chemists—if we were to call them either by the name
of the first authors who discovered them, or by the name of the substances
from which they were extracted—this would result in a confusion that even
someone with the most exceptional memory could not unravel. This method
might have been tolerable in chemistry's infancy and even perhaps as few as
twenty years ago when no more than thirty kinds of salts were known. But
now, when the number of salts increases every day, when each acid that is discovered often enriches chemistry with twenty-four new salts, and sometimes
with forty-eight when an acid has two degrees of oxygenation, a new method
is necessary, and this method is supplied by analogy: it is the one we adopted [184]
in the system of naming the acids. And since nature follows a single path, the
method will apply naturally to the nomenclature of neutral salts.

17.12 When we named the different kinds of acids, we distinguished the
particular acidifiable base of each acid from the acidifying principle, oxygen,
common to all. We expressed the property common to all of them by the
generic name "acid," and we then differentiated the acids by the name of the

---

17.11 *vitreous metallic compounds*: Such compounds were traditionally termed
"vitriols" because of their glassy (Latin *vitrum*) appearance.

acidifiable base which is peculiar to each. In this way we named oxygenated sulphur, phosphorus and carbon "sulphuric acid," "phosphoric acid" and "carbonic acid" respectively. And, finally, we thought we ought to indicate the different degrees of saturation of oxygen with a different ending for the same word. Thus we distinguished sulphurous acid from sulphuric acid, and phosphorous acid from phosphoric acid.

17.13 These principles, when applied to the nomenclature for neutral salts, required us to supply a common name for all salts produced through combination with the same acid, and then to differentiate them by the name of the salifiable base. So we used the name "sulphates" for all salts produced with sulphuric acid, "phosphates" for salts produced with phosphoric acid, and likewise for the others. We will therefore distinguish between "sulphate" of potash, "sulphate" of soda, "sulphate" of ammoniac, "sulphate" of lime, and "sulphate" of iron, etc.; and since there are twenty-four known bases, alkaline as well as earthy and metallic, there will be twenty-four kinds of "sulphates," as many "phosphates," and likewise for all the other acids. But since sulphur is capable of two degrees of oxygenation, where one measure of oxygen yields sulphurous acid and another, sulphuric acid, and since the neutral salts which these two acids form with different bases are not the same, having very different properties, it was necessary to distinguish them further by a particular ending. Consequently we have named the neutral salts formed by the less oxygenated acid "sulphites," "phosphites," and so on. Thus oxygenated sulphur will be capable of forming forty-eight neutral salts, namely, twenty-four "sulphates" and twenty-four "sulphites," and likewise with the other substances capable of two degrees of oxygenation.

17.14 It would be excessively tedious for readers to follow all of the details of this naming process. It suffices to have clearly presented the method. Once this has been grasped, it will be effortlessly applicable to every possible compound. When the name of the combustible, acidifiable substance is known, it will always be easy to remember the name of the acid this substance can form as well as the names of all the neutral salts that must be derived from it.

17.15 I will therefore confine myself to these elementary notions. But at the same time, in order to satisfy those who might be in need of greater details, in the second part of this treatise I will add tables which will provide a general summary, not only of all the neutral salts, but of all chemical compounds. There I will add several short explanations of the most simple and reliable ways to procure the different kinds of acids, and of the general properties of the neutral salts which result from them.

[185]

[186]

17.16 I do not deny that in order to complete this work it would have been necessary to add to these explanations particular remarks about each kind of salt, about its solubility in water and spirit of wine, about the ratio of acid and base which enter into its composition, about the quantity of water required to crystallize it, about the different degrees of saturation of which it is capable, and finally, about the degree of force with which the acid clings to [187] its base. Bergman, de Morveau, Kirwan, and several other famous chemists have begun this immense work. But it has as yet only modestly advanced, and the bases on which it rests are even now not rigorously accurate. Besides, such numerous details would not have been suitable in an elementary work, and the time needed to gather the materials and to complete the experiments would have delayed publication of this work by several years. This is a vast field, open to the zeal and activity of young chemists. But as I complete my task here, allow me to recommend to those who possess the courage to undertake this work to strive to do well rather than to do much; to ascertain, at the outset, the composition of acids by many precise experiments, before focusing on the composition of neutral salts. Every edifice destined to brave the ravages of time must be built upon solid foundations, and in the state chemistry has attained, its march would be impeded by building its progress upon experiments which are neither sufficiently precise nor sufficiently rigorous.

17.16 *spirit of wine*: Alcohol.

strive to do well rather than to do much: In addressing this advice to "young" chemists, does Lavoisier mean to imply that, although his own generation (Lavoisier was 44 when he wrote these words) has been granted a long and sweeping vision, from now on only patient, careful, and meticulous effort will make that vision a reality? If, indeed, *doing well* truly conflicts with *doing much*, Lavoisier's counsel has much to recommend it. But perhaps Lavoisier's own achievements testify that the very attentiveness and conscientiousness which characterize the *Treatise* may further one's accomplishment more than they restrain it.

Appendix

1. Pre-Revolutionary French Weights and Measures

Units of Length

point	$1/12^3$ *pied*	.188 mm or .074 in	
ligne	$1/12^2$ *pied*	2.256 mm or .089 in	corresponds to English *line*
pouce	$1/12$ *pied*	27.07 mm or 1.066 in	corresponds to English *inch*
pied du roi *or* pied		32.48 cm or 1.066 ft	corresponds to English foot; known in English as the *Paris foot*
toise	6 *pied*	6.394 ft	corresponds to English *fathom*

Units of (Liquid) Volume

demiard	1/4 *pinte*	238 ml or 1/2 US pint
chopine	1/2 *pinte*	476.1 ml or 1 US pint
pinte		952.1 ml or 2.01 US pint
quade	2 *pinte*	1.904 L or 1/2 US gal
quartaut	72 *pinte*	68.55 L or 36 US gal
pouce cube	1/48 *pinte*	19.84 ml or .671 US fl oz
pied cube	36 *pinte*	34 .28 L or 9.056 US gal

Units of Mass

prime	$1/24^3$ *once*	2.213 mg
grain	$1/24^2$ *once*	53.11 mg
gros	1/8 *once*	3.824 g or 0.135 US oz
once	1/16 *livre*	30.59 g or 1.079 US oz
livre		489.5 g or 1.079 US lb
quintal	100 *livre*	48.95 kg or 107.9 US lb

2. Conversion to Decimal Livres

Lavoisier expresses weights in *livres, once, gros,* and *grains* in accordance with the preceding table. We may therefore convert these quantities to decimal livres as follows:

livres	÷	1	=		livres
onces	÷	16	=		livres
gros	÷	128	=		livres
grains	÷	9216	=		livres
		TOTAL	=		decimal livres

For example, 2 livres, 9 onces, 3 gros and 17 grains of oxygen will amount to

2 livres	÷	1	=	2	livres
9 onces	÷	16	=	.563	livres
3 gros	÷	128	=	.023	livres
17 grains	÷	9216	=	.003	livres
		TOTAL	=	2.589	livres

3. Degrees of Temperature

The Réaumur temperature scale designates the freezing and boiling points of water as 0° and 80°, respectively. The chart displays it in comparison to the Celsius and Fahrenheit scales.

Glossary of Names and Terms

Academy: Refers to the French Academy of Sciences (*Académie des sciences*), a learned society founded in 1666 by Louis XIV. One of the earliest of such academies, it was at the forefront of scientific developments in Europe in the 17th and 18th centuries.

acetic acid: *See* radical vinegar.

acetous acid: *See* vinegar.

acid: "Having the taste of vinegar," from Latin *acidus*, "sour, sharp, tart." Lavoisier notes the parallel with Greek ὀξύς, "sharp."

aeriform: Similar to atmospheric air in having no fixed volume (unlike a liquid) and no fixed shape (unlike a solid).

affinity: Refers to the tendency of a substance to combine by chemical reaction with another substance. *See* Bergman; *see also* double affinities.

alembic: A complete distilling apparatus, consisting of a heated vessel (cucurbit) containing the substance to be distilled, an apparatus for condensing the vapors produced, and a receiver for collecting the liquid condensate. From Greek βῖκος, pot or vase.

alembroth, salt of: A substance supposed by some alchemists to be a universal solvent, later used as an antiseptic. Also called "salt of wisdom," "philosopher's salt."

algaroth: A white solid, formerly used in powdered form as an emetic.

alkali: A class of materials having properties similar to those of soda, which, being extracted from the herb kali, is the paradignmatic "al-kali." Lavoisier discusses the name in Chapter Sixteen. *See also* kali.

alum, alumina: Lavoisier's *alum* is an astringent mineral salt, *alumina* the name of its presumptive base, not yet isolated in Lavoisier's time. In 18th-century English, an alternate term for alumina was "pure clay."

amalgam: "Combination of mercury with another metal," from Latin *amalgama*, "alloy of mercury, especially with gold or silver"; perhaps an alteration of Latin *malagma* "poultice, plaster," or Greek *malagma*, "softening substance," from *malassein* "to soften." More generally, any blend or conglomeration.

aurora borealis: Northern Lights (lit. "northern dawn"). Auroras seen from sufficiently northern locations may be directly overhead, but viewed from lower latitudes they illuminate the poleward horizon as a greenish glow, or sometimes a faint red, as if the Sun were rising from that direction. The analogous phenomenon in the southern hemisphere is *aurora australis*.

barytes: A white or colorless mineral, used as a white pigment. From Greek βαρύς, heavy.

base: Modern readers know this term as the opposite of *acid*; but Lavoisier uses it in a much more general sense: as the foundation, so to speak, of a more complex substance. For Lavoisier, a "base" is always a base of something, not a definite nature in itself. *See also* Rouelle, Guillaume-François.

Becher, Johann Joachim (6 May 1635 – October 1682): A German physician, alchemist and scholar, best known for his development of the phlogiston theory of combustion.

Bergman, Torbern Olof (20 March 1735 – 8 July 1784): A Swedish chemist and mineralogist noted for his 1775 *Dissertation on Elective Attractions*, containing the largest chemical affinity tables ever published. *See also* affinity.

Berthollet, Claude Louis (9 December 1748 – 6 November 1822): A French chemist who in 1785 was first to determine the elemental composition of the gas ammonia.

Boerhaave, Herman (31 December 1668 – 23 September 1738): A Dutch botanist, chemist, and physician. He was the first physician to utilize the thermometer in clinical practice.

Boyle, Robert (25 January 1627 – 31 December 1691): A natural philosopher, chemist, physicist and inventor. With the help of his assistant Robert Hooke he constructed an improved air pump, with which he carried out experiments on vacuum and established the relation between pressure and volume of air ("Boyle's Law").

calcareous: Containing lime or limestone. *Chalk* was formerly known as "calcareous earth."

calcination: The process of thoroughly burning or roasting a mineral or metal, so as to consume or drive off all its volatile parts. Lavoisier describes calcination in Chapter Seven. *See also* calx.

calorimeter: Lavoisier's instrument for measuring a quantity of heat (caloric) by the amount of ice it is able to melt.

calx: A powder or friable substance produced by calcination of a mineral or metal. Sometimes denotes the calx *lime* in particular. Plural *calces*.

carbon: See charcoal.

Chaptal, Jean-Antoine (5 June 1756 – 30 July 1832): A distinguished French chemist, physician, agronomist, industrialist, statesman, educator and philanthropist. Chaptal's first scientific treatise, the *Éléments de chimie* (3 vols, Montpellier, 1790) introduced the term "nitrogen" in accordance with Lavoisier's system of chemical nomenclature. Nevertheless, Lavoisier's original term *azote* remains current in French.

charcoal: Lavoisier uses the term *charbon*, translated here as "charcoal," to refer both to a simple combustible substance that is a constituent of carbonic acid gas, and also to the fuel we call "charcoal" today, the residue of vegetable matter when decomposed at a high temperature in the absence of air.

colcothar: The brownish red oxide of iron which remains after the distillation of sulphuric acid from iron sulphate. Also called Crocus Martis, its finely powdered form constitutes jewelers' rouge.

compound: The present translation renders Lavoisier's *combinaison* as "compound" when it refers to the *product* of a chemical reaction and as "combination" when it refers to the *process* of chemical reaction.

Condillac, Étienne Bonnot (Abbé) de (30 September 1714 – 3 August 1780): French philosopher and friend of Rousseau, strongly influenced by Locke. Condillac's radical empiricism was popular in the second half of the 18th century in France. He is known, among other achievements, for his contributions to the philosophy of language and for a thought experiment, in defense of his empiricist epistemology, in which he demonstrated how a statue animated by a human soul could acquire all the ideas human beings possess through sensation.

crucifers: The family of flowering plants also known as the mustards or the cabbage family. The name, meaning "cross-bearing," refers to the four petals of mustard flowers, which resemble a cross. The family contains the cruciferous vegetables, including broccoli, cabbage, cauliflower, kale, collards, turnip, Chinese cabbage, radish and horseradish.

cucurbit: That part of a distilling apparatus which contains the material to be processed. From Latin *cucurbita*, gourd. *See also* alembic.

decompose: To separate a substance into its constituent parts. In Chapter Seven Lavoisier makes clear that since *caloric* is a constituent of gases, even elementary gases like oxygen may be said to be "decomposed."

de Fourcroy, Antoine François (15 June 1755 – 16 December 1809): A French chemist and physician who collaborated with Lavoisier, Guyton de Morveau, and Claude Berthollet on the *Méthode de nomenclature chimique*, a work that helped standardize chemical nomenclature.

de Luc, Jean-André (8 February 1727 – 7 November 1817): A Swiss geologist and meteorologist. De Luc studied the phenomena of boiling and freezing and ascertained that water exhibits its greatest density not at its freezing point but at a temperature several degrees higher.

de Morveau, Louis-Bernard Guyton (4 January 1737 – 2 January 1816): A French chemist and politician who, along with Lavoisier, Berthollet, and de Fourcroy, published the *Méthode de nomenclature chimique* in 1787.

deflagrate: To burst into flame and burn away with great rapidity.

deliquesce: To melt or become liquid by absorbing moisture from the atmosphere—a characteristic behavior of several salts.

Dippel's oil: Named after its inventor, Johann Conrad Dippel, this substance is a by-product of the destructive distillation of bones and is therefore sometimes known as *bone oil*. A dark, viscous, tar-like liquid with an unpleasant smell, its primary use was as an animal and insect repellent.

distillation: A process of separating the component substances from a liquid mixture by selective evaporation and condensation, made possible by differences in the volatility of the mixture's components. Distillation is properly a physical rather than a chemical process; but chemical reactions are sometimes unavoidable, as Lavoisier notes in Part III, Chapter Six of the complete Treatise.

docimasy: Close examination of a person or substance in order to determine nature, quality and characteristics. In ancient Greece, the process of evaluating aspirants for public office or citizenship. In metallurgy, the process of assaying metallic ores.

double affinities: A substance capable of uniting with two other substances is said to exhibit a double affinity.

dry way: *See* humid way.

earths: a loose category of dry, crumbly, odorless substances, generally mined from, or otherwise associated with, the earth. Earths were long regarded as irreducible, but Lavoisier knows several presumed earths to be metallic oxides.

effervescence: The escape of gas, in the form of bubbles, from a liquid. From Latin *effervescere*, "to boil up, boil over," from *ex-* "out" + *fervescere*, "begin to boil," from *fervere* "seethe, boil" (hence *brew*). *See also* ferment.

effloresce: "To flower out." In chemistry, the term denotes migration of a salt to the surface of a porous material, where it forms a coating.

element: Lavoisier uses the terms "element," "principle," and (less frequently) "simple substance" more or less interchangeably. The term "element" is usually preferred when Lavoisier wishes to emphasize that the substance is, so far as he can tell by means of analysis, irreducible; whereas the term "principle" is usually preferred when he wishes to emphasize the fact that the substance constitutes the essence of a particular type of compound.

empyreumatic: Consisting of, or having the odor of, burnt organic matter as a result of decomposition at high temperatures. From Greek ἐμπύρος, fiery.

-et: *See* -uret

ferment: As a verb, from Latin *fermentare*, "to leaven, cause to rise," from *fermentum*, "substance causing fermentation, leaven, drink made of fermented barley"; perhaps derived from *fervere*, "to boil, seethe"; hence *brew*. Lavoisier describes the process of fermentation in Chapter Thirteen. *See also* effervescence *and* yeast.

fixed air: So called by Joseph Black because it could be "fixed" (absorbed) by strong alkalis, renamed *carbonic acid* by Lavoisier.

fixed oils: They are "fixed" inasmuch as they do not readily evaporate. Note that this is a different sense of the term than its use in "fixed air." *See also* volatile.

flask: From Latin *flasco*, "container, bottle."

French thermometer: *See* Réaumur.

"fully formed": When one substance is a constituent of another substance, it is said to be "fully formed" when its own principles are actually combined rather than merely being present individually in the complex substance. For example, in paragraph 13.16 Lavoisier reports his changed opinion as to whether the hydrogen and oxygen in sugar were actually united in the form of water.

gallnuts: The hardened secretions which some plants (notably oak and sumac) produce when irritants are released by certain species of insect larvae. Oak gallnuts have a brown, globular, nutlike appearance.

gas: Matter in the state of an elastic aeriform fluid, "the last degree of saturation of some substance by caloric" (Chapter Four). The name, originated (in Dutch) by van Helmont, derives from Latin *chaos*.

gasometer: A massive apparatus designed by Lavoisier to deliver a measured volume of gas at constant pressure.

Gellert, Christlieb Ehregott (11 August 1713 – 18 May 1795): A German metallurgist, mineralogist, and educator. He composed influential textbooks on docimasy and chemistry, which Lavoisier knew through Baron d'Holbach's translations. In 1751 he published an elaborate table of chemical affinities drawing on the work of Geoffroy.

Gengembre, Philippe (1764–1838), a student of Lavoisier, obtained in 1783 an odiferous, flammable gas by heating phosphorus in a solution of potash and water. This gas was at first assumed to be phosphorus in a gaseous form; but Lavoisier recognised it as a combination of phosphorus with hydrogen and described it as *phosphure d'hydrogène* (phosphorated hydrogen gas).

Geoffroy, Étienne François (13 February 1672 – 6 January 1731): A professor of medicine and chemistry and a member of both the Royal Society and the Academy of Sciences, he played a major role in the transmission of scientific discoveries between England and France. He is known for the table of chemical affinities he published in 1718.

glasswort: *see* kali.

graphite: A crystalline form of carbon, formerly called *black lead* or *plumbago*; both these names arose from confusion with the similar-appearing lead ores, particularly galena. Abraham Gottlob Werner coined the name *graphite* ("writing stone") in 1789 in an attempt to clear up the confusion between molybdena, plumbago and black lead after Scheele proved in 1778 that they are different minerals.

humid way: In alchemical literature, the terms *humid way* and *dry way* refer primarily to methods for creating gold. For Lavoisier, however, they apply, respectively, to chemical operations that involve, or do not involve, water.

hyacinth (gemstone): A variety of zircon.

Ingenhousz, Jan (December 8, 1730 – September 7, 1799): A Dutch physiologist, biologist and chemist, best known for having discovered photosynthesis and the essential role played in it by light. He also discovered that plants, like animals, carry out respiration at the cellular level.

kali: A common seaside plant, also known as glasswort and saltwort; it has thick, fleshy, cylindrical leaves, each bearing a stout thorn. Its ashes yield an important ingredient used in making glass and soap. *See* soda.

kaolin: Also known as China clay, it differs from other clays in its ability to disperse in water. In 7th- and 8th-century China it was the principal ingredient in the manufacture of porcelain. The name *kaolin* derives from a Chinese phrase describing its place of discovery.

karabe: Amber. The name *karabe* derives from a Persian phrase meaning "attracting straws."

Kirwan, Richard (1 August 1733 – 22 June 1812): An Irish scientist, one of the last supporters of the theory of phlogiston. Kirwan was active in the fields of chemistry, meteorology, and geology.

Kunkel's phosphorus: Elemental "white" phosphorus. First produced by Hennig Brandt in 1669, it emits a faint glow when exposed to the oxygen in air and was accordingly named φωσφόρος, "light-bearer." Johann Kunkel (or Kunckel) afterwards produced it in commercial quantities, so that the material became widely known as "Kunkel's phosphorus." Phosphorus is highly flammable—a property utilized by Lavoisier in the experiments of Part I, Chapter Five.

Laplace, Pierre-Simon (23 March 1749 – 5 March 1827): A vastly influential French scientist whose work encompassed mathematics, statistics, physics and astronomy. From 1782 until 1784 Laplace and Lavoisier carried out joint investigations on the specific heats and the heats of combustion of various substances.

lime: An alkaline earth, the chief ingredient in masonry mortar. It is obtained by exposing limestone to a red heat; carbonic acid is driven off, and the brittle white solid that remains is pure lime, also called *quicklime*. Lime is strongly caustic and readily combines with water to form *slaked lime*, releasing considerable heat in the process.

limestone: A rock consisting chiefly of carbonate of lime; it yields lime when fired to red heat. *Marble* is limestone in crystalline form.

lixiviate: To extract a soluble constituent by washing and filtering. From Latin *lixivius*, "made into lye," from *lix*, "ashes, lye."

lute: A type of clay or cement used to stop holes, seal joints, etc. From Latin *lutum*, "mud, dirt, mire, clay."

Macquer, Pierre-Joseph (9 October 1718 – 15 February 1784): An influential French chemist opposed to Lavoisier's views. He published the *Dictionnaire de chymie* in 1766 and was also active in French industry, particularly the manufacture of porcelain.

magazine: A place for storing goods, especially military ammunition; from Arabic *khazana*, "to store up." (The English sense of "periodical journal" dates from 1731.)

marble: *See* limestone.

metal: From Latin *metallum*, "mine, quarry, mineral, what is got by mining," from Greek *metallon*, "metal, ore," probably related to *metallan*, "to seek after." Compare Greek *metalleutes*, "a miner," *metalleia* "searching for metals, mining."

Methodical Encyclopedia (*L'Encyclopédie méthodique ou par ordre de matières*): An expanded and updated version of the better known *L'Encyclopédie, ou dictionnaire raisonné des sciences, des arts et des métiers* edited by Diderot and d'Alembert. Many of the chemists to whom Lavoisier refers in the *Elementary Treatise* contributed to this Encyclopedia, which ran to over 200 volumes published over the course of fifty years, beginning in 1782.

Meusnier de la Place, Jean Baptiste Marie Charles (19 June 1754 – 13 June 1793): A French mathematician, engineer and Revolutionary general. Best known for Meusnier's Theorem on the curvature of surfaces, he worked with Lavoisier on the decomposition of water and evolution of hydrogen.

Monge, Gaspard (9 May 1746 – 28 July 1818): A French mathematician, best known for introducing what is now called orthographic projection of 3-dimensional curves. But Monge was also active in metallurgical and chemical researches that included the composition of nitrous acid, iron and steel.

neutral salt: A salt whose solution in water exhibits neither acidic nor alkaline properties.

nitre: Also called saltpeter, nitre is a white or colorless mineral which forms encrustations on the walls and ceilings of caves. It is also present in accumulations of guano.

noxious gas: Lavoisier uses *mofète* and *mofette*, which the present translation renders as "noxious gas," following Macquer's definition in the *Dictionnaire de chymie* to mean "unwholesome exhalations or vapors." But Lavoisier sometimes uses the term to refer specifically to the non-respirable part of the air, which he calls *azote gas* or, alternatively, *azotic gas.*

oil: In Chapter Ten Lavoisier describes oils as resulting from "hydrogen and charcoal uniting without the caloric bringing hydrogen into a gaseous state."

oxide: The first degree of oxygenation that does not form an acid. Lavoisier discusses this usage in Part I, Chapter Seven.

Papin's machine: Refers to Papin's "digester," a kind of pressure cooker designed to extract fat from bones, rendering them brittle and easily ground into bone meal.

Paracelsus (c. 1493–24 September 1541): A Swiss physician and alchemist, among the first to incorporate chemical knowledge in medical practice. His full name was Philippus Aureolus Theophrastus Bombastus von Hohenheim.

Peruvian bark: Formerly used as a treatment for malaria, since it contains quinine. Also called cincona bark, China bark, or Jesuits' bark.

phagedenic water: From Greek *phagedainikos*, "flesh-eating." A preparation from quicklime and corrosive sublimate, formerly used to remove fleshy growths.

plumbago: "Black lead," alternatively, any of several types of lead ore; from Latin *plumbum*, "lead." *But see* graphite.

pneumatic trough: A device originated by Stephen Hales in 1727 for collection of gases over mercury or water.

pneumato-chemical apparatus: A device that enables the separation of different gases in order to weigh them. Lavoisier describes this equipment in Part III, Chapter Six of the complete Treatise; the present volume offers an illustration in connection with paragraph 13.27.

pompholix: Zinc oxide, especially in the crude form produced by sublimation in a furnace.

potash: A water-soluble substance extracted from the ash produced when vegetable matter is burned in air. Lavoisier discusses it in Chapter Sixteen.

Pyrites: Refers to a group of minerals, but chiefly to the most common among them, *pyrite* or *iron pyrite* (fool's gold). Both names derive in part from its metallic, sparkling appearance; compare Greek πῦρ, fire.

quicklime: *See* lime.

radical: Adjective from Latin *radicalis*, "of or having roots," from Latin *radix* (genitive *radicis*), "root" (hence *radish*). As a noun, the root part of a word (linguistics); the root principle of a compound substance (chemistry).

radical vinegar: In Lavoisier's time, the common name for highly concentrated vinegar. Lavoisier assumed, pending confirmatory experiments, that it actually contained more oxygen than ordinary vinegar (which he called acetous acid), and therefore named it *acetic acid.* Today, highly concentrated vinegar is understood to be nothing other than vinegar itself; and both are now called *acetic acid.*

Réaumur, René Antoine (28 February 1683 – 17 October 1757): A French scientist who contributed to many different fields, especially the study of insects. In 1730 he introduced a scale of temperature based on the expansion rate of alcohol; but long before Lavoisier's time Réaumur's original scale was redefined according to the freezing and boiling points of water; it is styled "the French thermometer" by Lavoisier in Chapter One.

rectification: The process of purifying a distilled product by subjecting it to one or more subsequent distillations.

red heat: In Chapter Twelve Lavoisier writes: "Although it does not express a determinate degree of heat, it does, however, express a degree of heat much greater than that of boiling water."

resin: Any of a number of sticky, viscous substances secreted by plants for protection in response to injury; examples include pine gum, copal, and mastic. Amber is a fossilized resin.

retort: A vessel with a bent neck, from Latin *retorta*, literally "a thing bent or twisted," from past participle stem of Latin *retorquere*.

revivification: The restoration of a metal or other substance to a pure or uncombined state.

rock crystal: A clear, transparent variety of quartz.

Rouelle, Guillaume-François (16 September 1703 – 3 August 1770): The elder of two brothers who were both chemists and apothecaries. As Professor of Chemistry in the Royal Garden of Medicinal Plants, Rouelle gave public lectures for over twenty years, some of which Lavoisier attended. He is best known for his work on salts, and for introducing the term "base" in the sense of *a substance with which acids react to form neutral salts*.

reverberatory furnace: A type of smelting furnace in which the ore is separated from the fuel; the heat is then made to *reberverate* onto the material being processed (an old sense of *reverberate* means to turn or bend back).

salt: In Chapter Sixteen Lavoisier restricts the term "salt" to "composites formed by joining a simple oxygenated substance to a given base."

saltwort: *See* kali.

Saussure, Nicolas-Théodore de (14 October 1767 – 18 April 1845): A Swiss chemist and student of plant physiology who made discoveries in plant-derived chemicals and in the study of photosynthesis.

slaked lime: From *slackened*, that is, rendered less vigorous. *See* lime.

spirit of wine: Lavoisier mentions this term, in addition to "eau-de-vie" and "aqua vitae," as referring to pure alcohol; although these terms were, and still are, often used to refer particularly to brandy (distilled wine), a liquor that is not usually pure.

soda: A substance extracted from the ashes of the herb kali. Both "kali" and "soda" derive from Arabic names for the same herb; but in 18th-century usage, "soda" (French *soude*) denotes not the herb but the substance obtained from it.

Stahl, Georg Ernst (22 October 1659 – 24 May 1734): A German chemist, physician and philosopher. He was a supporter of vitalism, and until the late 18th century his writings on phlogiston provided the accepted explanation of combustion.

sublimate: As a verb, the modern meaning is to pass directly from the solid to the gaseous state; but an older sense means to undergo distillation. As a noun, it formerly referred to a substance produced by distillation, as in "corrosive sublimate" (sublimate of mercury).

sulphuric ether: A volatile and flammable liquid, made by distilling ethanol with sulphuric acid, and widely used as a solvent. It is the "ether" formerly (from about the middle of the 19th century) used in surgery as a general anaesthetic; but its anesthetic properties were known to Paracelsus in the 16th century.

Thouret, Michel Augustin: A French physician and reformer, appointed as the first director of the new School of Medicine at Paris in 1794.

tubulated: A retort or similar vessel is said to be tubulated if it is furnished with a stoppered opening through which substances may be introduced.

turpeth mineral: A lemon-yellow alkaline sulphate of mercury, obtained from common sulphate by washing with hot water. Also prepared from the leaves of the turpeth plant, it was formerly used as a cathartic.

-uret (suffix): The ending given to compounds of non-oxygenated substances (Chapter Ten).

van Helmont, Jan Baptist (12 January 1580 – 30 December 1644): A Flemish chemist, physiologist, and physician. He coined the word "gas" and was the first to recognize that there are gases distinct in kind from atmospheric air.

Vandermonde, Alexandre-Théophile (28 February 1735 – 1 January 1796): A musician, mathematician and chemist. He performed experiments with Lavoisier on the great frost of 1776, and published two papers on the manufacture of steel with Monge and Berthollet.

vinegar: A dilute acidic liquid produced by fermentation of wine or other alcoholic liquids. In Lavoisier's time it was also known as "acetous acid." Although the "-ous" suffix in the common term was a simple adjectival ending carrying no chemical significance, it fitted nicely with the systematic significance (i.e., indicating the state of oxidation) which Lavoisier would later give to the suffixes "-ous" and "-ic". Today, vinegar is called *acetic*, not *acetous*, acid. *See also* radical vinegar.

vitriol: Any of a number metallic compounds having a glassy (Latin *vitrum*) appearance. In Chapter Six Lavoisier recounts the connection between vitriol of iron ("green vitriol") and sulphuric acid; similar experiments would indicate that the common element in "vitriols" is sulphur.

volatile: "Fine or light," also "evaporating rapidly," from Latin *volatilis*, "fleeting, transitory; swift, rapid; flying, winged," from past participle stem of *volare*, "to fly." In Middle English, *volatiles* denoted birds, butterflies, and other winged creatures.

water of crystallization: Refers to the water that is necessary for the formation of some crystals, and which is incorporated in the crystal structure.

worm: A long spiral or coiled tube. In a distillation apparatus, it is the part in which the vapors are condensed. *See also* alembic.

yeast: The name comes from Old English *gist* or *gyst*, and from the Indo-European root *yes-*, meaning "boil", "foam", or "bubble." In Chapter Thirteen Lavoisier describes yeast as initiating fermentation by disturbing the equilibrium of the components of sugar. Not until 1857 would it be proved (by Louis Pasteur) that yeast's role in fermentation was that of a living organism. *See also* ferment.

Index